Advances in Chromatography

Advances in Chromatography

Volume 54

Edited by
Eli Grushka
Nelu Grinberg

CRC Press
Taylor & Francis Group
Boca Raton   London   New York

CRC Press is an imprint of the
Taylor & Francis Group, an **informa** business

CRC Press
Taylor & Francis Group
6000 Broken Sound Parkway NW, Suite 300
Boca Raton, FL 33487-2742

First issued in paperback 2022

© 2018 by Taylor & Francis Group, LLC
CRC Press is an imprint of Taylor & Francis Group, an Informa business

No claim to original U.S. Government works

ISBN 13: 978-1-03-240213-0 (pbk)
ISBN 13: 978-1-138-05595-7 (hbk)
ISBN 13: 978-1-315-11637-2 (ebk)

DOI: 10.1201/9781315116372

Visit the Taylor & Francis Web site at
http://www.taylorandfrancis.com

and the CRC Press Web site at
http://www.crcpress.com

Contents

Contributors

Jelena Čolović
Department of Drug Analysis
Faculty of Pharmacy
University of Belgrade
Belgrade, Serbia

Victor David
Department of Analytical Chemistry
Faculty of Chemistry
University of Bucharest
Bucharest, Romania

Slavica Erić
Department of Pharmaceutical
  Chemistry
Faculty of Pharmacy
University of Belgrade
Belgrade, Serbia

Ali Fouad
Chirality Program
Faculty of ESTEM
University of Canberra
Australian Capital Territory, Australia

Ashraf Ghanem
Chirality Program
Faculty of ESTEM
University of Canberra
Australian Capital Territory, Australia

Nelu Grinberg
Boehringer Ingelheim Pharmaceuticals
Ridgefield, Connecticut

Marko Kalinić
Department of Pharmaceutical
  Chemistry
Faculty of Pharmacy
University of Belgrade
Belgrade, Serbia

Ingela Lanekoff
Department of Chemistry – BMC
Uppsala University
Uppsala, Sweden

Julia Laskin
Physical Sciences Division
Pacific Northwest National Laboratory
Richland, Washington

Anđelija Malenović
Department of Drug Analysis
Faculty of Pharmacy
University of Belgrade
Belgrade, Serbia

Serban C. Moldoveanu
R. J. Reynolds Tobacco Co.
Winston-Salem, North Carolina

Ana Vemić
Department of Quality Assurance and
  Control
Slaviamed Ltd.
Belgrade, Serbia

1 Recent Progress in Fundamental Understanding and Practice of Chaotropic Chromatography

Rationalizing the Effects of Analytes' Structure with Pharmaceutical Applications

Ana Vemić, Marko Kalinić, Jelena Čolović,
Slavica Erić, and Anđelija Malenović

CONTENTS

1.1 INTRODUCTION

Analysis of basic compounds by reversed-phase high-performance liquid chromatography (RP-HPLC) is often affected by a number of issues relating primarily to the protonation of the analyte. Majority of drugs, for example, that contain basic ionizable groups have pK_a values more than 6 and most frequently around 9 (Manallack 2009). This implies that completely suppressing the ionization of most basic analytes, to induce adequate form for RP-HPLC analysis, requires adjusting mobile phase pH to values around 8–11, which lie outside the range suitable for the typical chromatographic bed of silica-based stationary phases.

When suppressing analyte ionization is not a viable approach in attaining acceptable retention behavior, other chromatographic modes need to be considered. One of the better established approaches in this respect is ion-interaction chromatography, often interchangeably referred to as ion-pair chromatography. This mode is an intermediate between reversed phase and ion-exchange chromatography and employs characteristic eluents containing ion-interaction agents (IIA). These additives are typically amphiphilic or liophilic in nature and have a tendency to accumulate onto the hydrophobic surface of the stationary phase, which leads to the formation of a pseudo ion-exchange surface. Electrostatic interactions between charged analytes and the oppositely charged IIA in both pseudo ion-exchange surface and the mobile phase modulate the chromatographic behavior in such systems so that acceptable retention, peak symmetry, and separation efficiency can be achieved (Cecchi 2009).

Amphiphilic additives have been in use as IIAs in chromatography for almost four decades, and efforts to provide a mechanistic explanation of their effects are roughly old (Eksborg et al. 1973; Knox and Laird 1976; Horvath et al. 1977; Kraak et al. 1977; Tilly Melin et al. 1979; Bidlingmeyer et al. 1979). Amphiphilic ions usually contain a charged group attached to a long alkyl chain. In chromatographic systems, at the interface between the hydrophobic stationary phase surface and the eluent, these ions become specifically oriented so that their charged heads project toward the bulk eluent, whereas the hydrophobic alkyl chain is adsorbed on the stationary phase. The application of these agents to a reversed-phase column usually results in it permanently retaining an ion-exchange character, which can compromise the reproducibility and robustness of separation, also clearly limiting the column's usability and lifetime (Makarov et al. 2008). This shortcoming has provided an incentive to investigate new potential IIAs, leading to the recognition that many simple inorganic and organic liophilic ions can serve as an alternative to classical amphiphilic agents. These small liophilic ions exhibit the same beneficial effects on the retention profile of charged analytes in RP-HPLC systems, but can easily be dissolved in the mobile phase, making their impact on initial column properties reversible. Liophilic ions are usually spherical, characterized by significant charge delocalization, symmetry, and large overall electron density. These ions are often referred to as chaotropic ions due to their ability to impair the solvation shell around charged analytes thereby introducing *chaos* in the structure of the ionic solution. Consequently, ion-interaction chromatography with chaotropic ions can more simply be referred to as *chaotropic chromatography*.

The aim of this review is to provide the reader with an update on the most recent developments in fundamental understanding of retention mechanisms in chaotropic chromatography and in the wide range of its practical applications in drug analysis. A particular emphasis is placed on novel efforts to model the influence of analytes' structure on its retention behavior in these chromatographic systems. For the convenience of the reader, in Sections 1.2 through 1.4, we open this discussion by providing a concise overview of the theoretical and practical foundations of chaotropic chromatography. Section 1.2 gives a brief phenomenological description of the putative retention mechanisms in chaotropic chromatography, with emphasis on how chaotropic agents behave in an RP-HPLC system, and how they, in turn, modulate the retention behavior of charged analytes. An overview of the equilibria that exist in the resulting complex chromatographic system is presented. Section 1.3 gives a critical overview of the theoretical approaches that can be used to model the retention behavior in chaotropic chromatography. Finally, Section 1.4 summarizes key empirical findings related to how mobile phase composition, choice of the chaotropic agent, and the chromatographic column affect retention phenomena. Building on these foundations, Section 1.5 provides a detailed review of emerging attempts to describe and rationalize the influence of analytes' structure on its retention in chaotropic chromatography through quantitative structure–retention relationship (QSRR) modeling. Section 1.6 turns to practical applications in the pharmaceutical field, providing several examples of the utility of chaotropic chromatography in the analysis of basic drugs, and possibilities to tailor separations in these systems via novel experimental design—design of experiment (DoE) approach. Finally, Section 1.7 concludes this discussion with remarks on outstanding challenges in the field and opportunities for further research efforts.

1.2 RETENTION MECHANISMS IN REVERSED-PHASE LIQUID CHROMATOGRAPHY WITH CHAOTROPIC ADDITIVES

Retention mechanisms in ion-interaction chromatography have been the subject of much debate in the literature, with several opposing views on key forces that govern retention in these systems. Several excellent reviews are available on the topic (Cecchi 2008; Ståhlberg 1999; Cecchi 2015), and the interested reader is referred to these publications for a more complete coverage. Here, we aim to provide only a brief outline of the most relevant concepts, to serve as basis for subsequent discussions.

As illustrated in Figure 1.1, a typical chromatographic system with IIAs can be described through a series of equilibria and the corresponding stoichiometric equilibrium constants, among which the most relevant are (Cecchi 2009):

1. Adsorption of the IIA (H) to surface ligand sites (L) of the hydrophobic stationary phase, leading to the formation of a charged surface site (LH)
2. Adsorption of the analyte—eluite (E) to the stationary phase surface
3. Ion pairing in the stationary phase (LEH)
4. Ion pairing in the mobile phase (EH)
5. Displacement of H by E

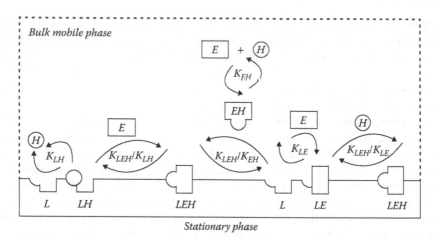

FIGURE 1.1 A schematic representation of the putative equilibria underlying the retention mechanisms in ion-interaction chromatography.

The earliest interpretations of the retention mechanism in ion-pairing chromatography pioneered the idea that the formation of a neutral ion pair (*EH*) in the mobile phase is a key event leading to prolonged retention of the charged analyte (Wittmer et al. 1975; Knox and Laird 1976; Horvath et al. 1977). The diminished polarity and increased effective lipophilicity of *EH* were expected to favor its adsorption on ligand sites of the hydrophobic stationary phase (*LEH*) thus extending its retention time. Other authors proposed an alternative hypothesis, whereby adsorption of the IIA to the stationary phase generates charged sites (*LH*) where dynamic ion exchange can occur (Kraak et al. 1977; Kissinger 1977; Ghaemi and Wall 1979)—leading also to prolonged retention of the analyte in the stationary phase via the formation of an adsorbed neutral ion pair (*LEH*). Although at first, these two mechanisms appear distinct, they are in fact thermodynamically identical (Knox and Hartwick 1981) as the initial and the final states are the same, while only the order of intermediate steps is different. It is important to observe that ion-pair formation is pertinent to all stoichiometric descriptions of retention mechanisms involving chromatographic systems with IIAs. Understanding the effectiveness of chaotropic agents as IIAs thus requires considering specific ion effects, which putatively favor ion pairing with organic solutes.

Specific ion effects occur both in bulk solution and at interfaces. They represent a set of phenomena that have puzzled physical chemists for decades and continue to be an active area of research (Parsons et al. 2011). These specific ion effects are often referred to as Hofmeister effects, owing to a series of papers published at the end of the nineteenth century by a German pharmacologist Franz Hofmeister, who first described the differential efficiency of various ions in precipitating egg yolk proteins, ranking them in what has become known as the Hofmeister series (Lo Nostro and Ninham 2012). Consideration of Hofmeister effects in bulk dilute solutions is often coupled to considering perturbations in the ordered structure of water, whereby it is possible to differentiate between *structure makers* or *kosmotrope* ions and *structure breakers* or *chaotrope* ions. The rank of an ion in the Hofmeister series is essentially related to its effects near a surface (of a protein), which are not necessarily correlated to its effects in bulk solution. Nonetheless, it is often taken that the Hofmeister series is also a measure of how chaotropic or kosmotropic an ion is (Marcus 2009). For common anions, the series can be written as: $CO_3^{2-} < SO_4^{2-} < S_2O_3^{2-} < H_2PO_4^- < OH^- < F^- < HCOO^- < CH_3COO^- < Cl^- < Br^- < NO_3^- < I^- < ClO_4^- < SCN^-$, where the thiocyanate ion would be the most chaotropic, and carbonate the least chaotropic ion in the series. Anions, which are commonly part of the mobile phase in chaotropic chromatography can likewise be ranked in order of increasing chaotropicity (Cecchi and Passamonti 2009): $H_2PO_4^- < HCOO^- < CH_3SO_3^- < Cl^- < NO_3^- < CF_3COO^- < BF_4^- < ClO_4^- < PF_6^-$. A rigorous description of the physical basis that determines an ion's rank in the Hofmeister series has proven somewhat elusive (Parsons et al. 2011; Lo Nostro and Ninham 2012), but it is interesting to note that the rank can be related to a number of physical observables such as (Collins and Washabaugh 1985): viscosity B coefficients in the empirical Jones–Dole equation, surface potential difference and surface tension at the air–water interface, polymer swelling, protein solubility and denaturation temperature, protein aggregation propensity, solubility and critical micellar concentration of organic solutes, and so on.

The ability of chaotropic ions to perturb the ordered structure of water is an important concept in explaining the effectiveness of these agents in forming ion pairs with charged organic solutes. Solvent effects have long been known to affect ion-pair formation, but these were typically rationalized only in terms of electrostatic screening (Marcus and Hefter 2006). It was Diamond who first pioneered the idea that ion pairing can be driven by perturbations in water structure (Diamond 1963); chaotropes are poorly solvated and so are amphiphilic ions (charged analytes or classical IIAs), which typically induce formation of cage-like water structures around them, incurring an entropic penalty. Their ion pairing can thus restore water structure closer to its original state, making it energetically favored (Marcus and Hefter 2006). Importance of the concept of hydrophobic ion pairing in chromatography has been recognized relatively early on (Tomlinson et al. 1978), and it provides a basis for understanding some of the specific effects chaotropes exert.

However, neither the ion pairing nor the dynamic ion-exchange model can fully account for all experimental observations in ion-interaction chromatography. A more comprehensive description of the retention mechanism, incorporating some aspects from both previously outlined models, was put forth by Bidlingmeyer and coworkers (Bidlingmeyer et al. 1979; Bidlingmeyer 1980), who also first noted on the importance of considering electric double-layer formation in these chromatographic systems. Like amphiphilic IIAs, chaotropes are also adsorbophilic species, and the rank of an ion in the Hofmeister series interestingly coincides with its tendency to accumulate on the surface of the stationary phase in RP-HPLC, which can be quantified via adsorption isotherms (Kazakevich et al. 2005; Kazakevich and Snow 2006). This adsorbophilicity of the chaotropic agent has major implications for its efficiency in extending the retention of oppositely charged analytes. More specifically, under the assumption that the chaotropic ion is much more adsorbophilic compared to its counterion, it can be expected that the surface of the stationary will become charged due to the adsorbed layer of the chaotropic IIA. Oppositely charged ions from the eluent are then attracted to the charged surface, leading to the formation of an electric double layer. The analyte can interact with the electric double layer through dynamic ion exchange and/or by becoming adsorbed on the surface, due to the electrostatic potential difference that develops (Cantwell 1984; Liu and Cantwell 1991).

The magnitude of the developed potential (ψ) can be calculated by taking into account the Gouy–Chapman description of the electric double layer or the version of this theory modified by Stern (Cecchi 2009). The former models the double layer in a way that it consists of a primary layer of adsorbed ions and a secondary diffuse layer of counterions, both in dynamic equilibrium, with the electrostatic potential decaying exponentially at distance from the stationary phase surface (Figure 1.2a). The latter version by Stern describes two planes: the inner Helmholtz plane, which is defined by the layer of adsorbed ions, and the outer Helmholtz plane, which is located at the smallest distance that oppositely charged ions can approach the surface. At distances beyond the outer Helmholtz plane, a third, diffuse layer exists. The potential decays linearly in the compact layer between the two Helmholtz planes (from ψ^0 to ψ^{OHP}) and then exponentially in the diffuse layer (Figure 1.2b). Both of these models have alternatively been used in theoretical modeling of retention behavior in chaotropic chromatography, as discussed in Section 1.3.

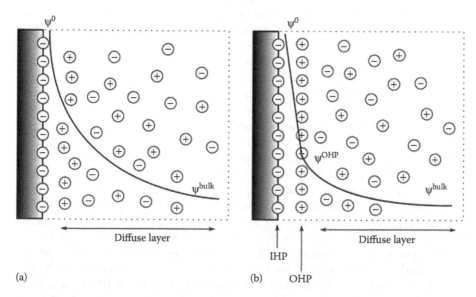

FIGURE 1.2 Schematic representations of structure of the electric double layer according to (a) the Gouy–Chapman model; (b) the Stern–Gouy–Chapman model. (*Abbreviations*: IHP—inner Helmholtz plane; OHP—outer Helmholtz plane; ψ—electrostatic potential.)

1.3   THEORETICAL MODELS OF ION-INTERACTION CHROMATOGRAPHY

Initial efforts to place experimental observations from ion-interaction chromatography into a consistent theoretical framework were largely based on a stoichiometric approach (Horvath et al. 1977; Knox and Hartwick 1981; Hung and Taylor 1980; Bruzzoniti et al. 1996; Sarzanini et al. 1996), considering the equilibria outlined in Figure 1.1. These models relied alternatively on the ion pairing and dynamic ion-exchange descriptions of the retention mechanism, with some theoretical models incorporating certain elements of both. However, as summarized by Cecchi (2015), stoichiometric models are unable to account for several experimentally observed phenomena, namely:

- Existence of maxima in the retention factor versus IIA concentration plots
- Retention decrease with increasing IIA concentration for analytes in the same charge status as the agent
- The slight influence of IIA concentration on the retention of neutral analytes

The principal shortcoming of stoichiometric models is the fact that they neglect the formation of an electric double layer on the surface of the stationary phase. In presence of an electrostatic potential difference at the interface between stationary and mobile phases, constants of stoichiometric models are essentially not constant, rendering the resulting theoretical description physically nonrigorous (Cecchi 2009).

The first nonstoichiometric models of retention in ion-interaction chromatography were based on the idea that electrostatic interactions with the surface double layer present the major driving force modulating the retention of charged analytes; consequently, they are commonly referred to as pure electrostatic models. Bartha and Ståhlberg (1994) developed a model based on the Gouy–Chapman description of the electric double layer, hypothesizing that the free energy of adsorption of the analyte onto the stationary phase can be broken down into two components: one reflecting the analytes' inherent hydrophobicity and the other corresponding to the electrostatic interaction with the surface double layer. Considering how the former effectively governs retention in absence of IIA, and the latter is proportional to the magnitude of the surface potential, these ideas are clearly manifested in the final form of the proposed model:

$$k = k_0 e^{\frac{-z_E F \psi^0}{RT}} \tag{1.1}$$

where:
z_E is the charge of the analyte
F is the Faraday constant
ψ^0 is the surface electrostatic potential
R is the gas constant
T is the absolute temperature

The authors proposed calculating the potential via a linearized expression of the Poisson–Boltzmann equation, assuming approximately planar surface geometry. The surface excess of the IIA was determined using a linearized, potential-modified Langmuir isotherm. Cantwell (1984) proposed a more complex model, based on the Stern–Gouy–Chapman description of the electric double layer. Other notable electrostatic models have been developed by Weber et al. (Weber and Orr 1985; Weber 1989) and Deelder et al. (Deelder and van den Berg 1981), though these have been applied less frequently. All electrostatic models include certain approximations in solving the Poisson–Boltzmann equation, in treating the effects of analytes' adsorption on the surface potential, and in evaluating the surface excess of the IIA (Chen et al. 1993; Cecchi 2009). However, what can be considered a more fundamental issue with these models is the fact that they neglect ion-pairing phenomena that occur between organic solutes and IIAs, and whose relevance is supported by substantial experimental evidence (Dai et al. 2005; Dai and Carr 2005; Fini et al. 1999; Takayanagi et al. 1997; Steiner et al. 2005).

Arguably the most comprehensive theoretical model put forth thus far has been proposed by Cecchi et al. (2001b). This model is based on what the authors refer to as an *extended thermodynamic approach* and aims to overcome the limitations of both stoichiometric and pure electrostatic models. Being the first proposed model for systems with classical amphiphilic IIAs, this model was also shown to be applicable for describing retention behavior in chaotropic chromatography (Cecchi and Passamonti 2009), as the principal differences are limited only to the comparatively low adsorbophilicity of IIAs used in the latter systems. The extended thermodynamic approach encompasses the most relevant equilibria from the stoichiometric description of the chromatographic system (as discussed in Section 1.2) but introduces thermodynamic equilibrium constants. Further, to account for the surface potential difference, those equilibrium constants corresponding to the interaction of charged species with the stationary phase are modified by a term accounting for the potential. This introduces the need to quantify the potential; within the framework proposed by the authors, the surface excess of the IIA can be modeled as a function of its mobile phase concentration using the Freundlich adsorption equation (Cecchi 2005):

$$[LH] = a \cdot [H]^b \qquad (1.2)$$

where:

[LH] is the surface excess of the IIA

[H] is its mobile phase concentration, with a and b representing fitting parameters, which can be experimentally evaluated

Combined with the Gouy–Chapman description of the electric double layer and the Poisson–Boltzmann equation for planar surface geometry, the potential can be calculated as originally proposed by Ståhlberg (1986):

$$\Psi^\circ = \frac{2RT}{F} \ln \left\{ \frac{[LH] \cdot |z_H| F}{\left(8\varepsilon_0 \varepsilon_r RT \sum_i c_{0i} \right)^{0.5}} + \left[\frac{\left([LH] \cdot z_H F \right)^2}{8\varepsilon_0 \varepsilon_r RT \sum_i c_{0i}} + 1 \right]^{0.5} \right\} \qquad (1.3)$$

where:

z_H is the charge of the adsorbophilic IIA ion

ε_0 is the electric permittivity of vacuum

ε_r is the relative permittivity of the mobile phase

Σc_{0i} is the mobile phase concentration of singly charged ions

By combining Equations 1.2 and 1.3, with equations for the relevant equilibria, the following final form of the model can be obtained:

$$k = \frac{c_1 \left\{ a[H]^b f + \left[\left(a[H]^b f \right)^2 + 1 \right]^{0.5} \right\}^{\pm 2|z_E|} + c_2 [H]}{\left(1 + c_3 [H] \right) \left\{ 1 + c_4 [H] \left\{ a[H]^b f + \left[\left(a[H]^b f \right)^2 + 1 \right]^{0.5} \right\}^{\left(-2|z_H| \right)} \right\}} \quad (1.4)$$

where:

$$f = \frac{|z_H| F}{\left(8\varepsilon_0 \varepsilon_r RT \sum_i c_{0i} \right)^{0.5}} \quad (1.5)$$

can be evaluated from mobile phase composition and experimental conditions. The fitting parameters of Equation 1.4 (c_1–c_4) have a clear physical meaning:

$$c_1 = \Phi[L]_T K_{LE} \frac{\gamma_L \gamma_E}{\gamma_{LE}} \quad (1.6)$$

$$c_2 = \Phi[L]_T K_{EHL} \frac{\gamma_E \gamma_H \gamma_L}{\gamma_{EHL}} \quad (1.7)$$

$$c_3 = K_{EH} \frac{\gamma_E \gamma_H}{\gamma_{EH}} \quad (1.8)$$

$$c_4 = K_{LH} \frac{\gamma_L \gamma_H}{\gamma_{LH}} \quad (1.9)$$

It should be noted that parameter c_1 is generally not considered adjustable since it equals k_0, the retention factor of the analyte without chaotropic additive in the eluent, which can be experimentally obtained. Parameters c_2, c_3, and c_4 are related to the thermodynamic equilibrium constants for the ion-pair formation in the stationary phase, for the ion-pair formation in the eluent, and for the adsorption of the ion-interaction reagent onto the stationary phase, respectively. Relatively low adsorbophilicity of common chaotropic salts means the last term can usually be omitted from the fitting procedure.

The extended thermodynamic approach has been criticized by Ståhlberg (2010) on grounds of it being *too complicated for practical use*, with the large number of adjustable constants suggesting that it would *fit numerically well to a large body of experimental retention data*; the author also argued that there was not a *single set of experimental retention data that is analyzed with the model.* Contrary to this opinion and as will be discussed in the subsequent sections, our group's results suggest that the extended thermodynamic approach is readily applicable and can further provide useful insights into the key phenomena governing retention.

1.4 FACTORS GOVERNING THE RETENTION IN CHAOTROPIC CHROMATOGRAPHY

1.4.1 EFFECT OF CHAOTROPIC ANION TYPE AND CONCENTRATION

Choice of the chaotropic salt and its mobile phase concentration have major effects on the retention behavior of protonated basic analytes, and this has been the subject of numerous chromatographic studies (LoBrutto et al. 2001a, 2001b; Jones et al. 2002; Roberts et al. 2002; Hashem and Jira 2006; Flieger 2006, 2007; Flieger and Świeboda 2008; Vemić et al. 2013, 2014, 2015a, 2015b; Čolović et al. 2015). The key findings of these investigations can be summarized as follows:

- Magnitude of the increase in basic analytes' retention increases with the rank of the chaotropic anion in the Hofmeister series—increasing chaotropicity of the IIA favors analytes' retention.
- Plots of retention factors versus chaotrope concentration typically follow a characteristic trend (Figure 1.3), with plateaus above a certain concentration of the chaotropic anion.
- Analytes' *chaotropic sensitivity* is proportional to its hydrophobicity—solutes of low hydrophobicity are far less affected by the addition of a chaotropic agent than their highly hydrophobic counterparts, which typically experience a very pronounced increase in retention. This is illustrated in Figure 1.3, comparing the retention behavior of tramadol ($ClogP = 2.45$), mianserine ($ClogP = 3.83$), and amitriptyline ($ClogP = 4.81$).

These experimental observations can be rationalized in terms of the ion-pairing mechanism of retention (Hashem and Jira 2006; Flieger 2007): the more chaotropic an ion, the more easily it disrupts analytes' solvation layer, leading to the formation of stable ion pairs; and the more hydrophobic the solute is, the longer will its neutral ion pair be retained in the stationary phase. However, considering ion pairing alone is not sufficient to explain all the differential effects of various chaotropic salts used in different concentrations. Specifically, if disruption of analytes' solvation was the only phenomenon involved, identical retention plateaus would be reached for a single analyte and different chaotropic salts, and this is clearly not consistent with experimental findings. To fully account for differential effects of various chaotropic anions, their relative adsorbophilicity needs to be considered (Kazakevich et al. 2005; Makarov et al. 2008). Namely, as illustrated in Figure 1.4, increasing

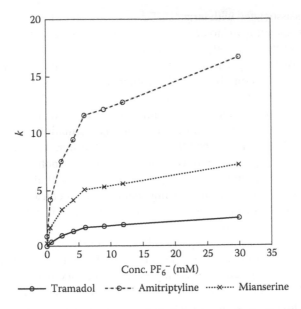

FIGURE 1.3 Plots illustrating the characteristic dependence of retention factors on the concentration of the chaotropic agent in the mobile phase. Data collected on column Luna C18 150 × 4.6 mm 5 μm (Phenomenex, CA, USA); mobile phase: acetonitrile-water solution of $NaPF_6$ (40:60% v/v); mobile phase pH 3; flow rate = 1 mL/min; λ = 254 nm.

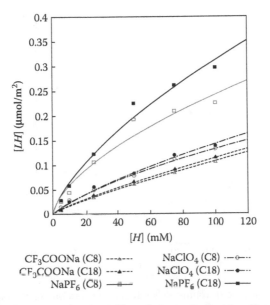

FIGURE 1.4 Differences in adsorbophility of three chaotropic salts, evaluated by the breakthrough method of frontal analysis in chromatographic systems with columns of differing hydrophobicity. Experimental data were fitted into the Freundlich adsorption equation to produce the shown plots.

chaotropicity is associated with larger surface excess of the chaotrope on the stationary phase, corresponding to larger surface potentials, and enhanced effects on retention of oppositely charged analytes (Vemić et al. 2015a).

Net effects of varying chaotrope type and concentration on analytes' retention are therefore the combined result of changes in ion pairing and magnitude of the surface potential—a fact that also accounts for emergence of plateaus in retention factor *versus* chaotrope concentration plots. Namely, it is important to emphasize that, contrary to initial interpretations, formation of ion pairs in the mobile phase essentially shortens the analytes' retention. This was nicely illustrated with the example of ropinirole, whose retention behavior was modeled within the extended thermodynamic framework by Vemić et al. (2014). As can be seen in Figure 1.5, if effects of increasing chaotrope concentration are considered only in terms of the surface potential, a linear increase in retention factors of ropinirole would be expected.

On the other hand, if only ion pairing is taken into account, a linear decline in retention would be expected; this is due to the fact that, with this particular analyte and chromatographic system, ion pairing in the stationary phase was found to be negligible, with only the c_3 term of the model appearing statistically significant. It is only when both effects of increasing chaotrope concentration are considered that one obtains a very accurate description of the experimentally observed retention behavior.

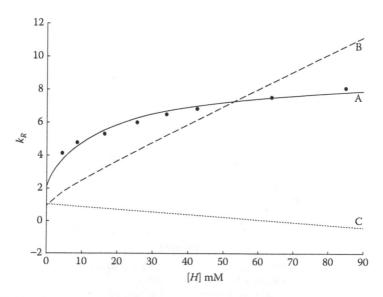

FIGURE 1.5 Dependence of ropinirole's retention factors on concentration of $NaClO_4$ as the chaotropic agent. Dots represented experimentally observed data. Curve A was obtained by fitting experimental data to Equation 1.4, whereas curves B and C represent contributions of electrostatic interactions and ion pairing in the eluent, respectively. (From Vemić, A. et al., *Talanta*, 123, 122–127, 2014.)

1.4.2   EFFECT OF ORGANIC MODIFIER TYPE AND ITS MOBILE PHASE FRACTION

Choice of the organic modifier and its mobile phase fraction have complex effects on the overall retention behavior of analytes in chaotropic chromatography, since these factors influence not only eluent strength but also affect the stationary phase surface *microenvironment* and adsorption of the chaotropic agent. Namely, commonly employed organic modifiers—methanol, acetonitrile, and tetrahydrofurane (THF)—are known to exhibit differential adsorption behavior on hydrophobic stationary phases, whereas methanol forms monolayers; both acetonitrile and THF are capable of forming adsorbed layers more than 10 Å thick (Kazakevich et al. 2001).

Stationary phase surface adsorption of acetonitrile strongly favors adsorption of chaotropes, with this effect being more pronounced with more chaotropic anions. Such an enhancement is, conversely, only very moderate in the case of methanol. However, increasing acetonitrile percentage in the mobile phase above a certain threshold leads to opposite effects, as the stationary phase is approaching saturation, and as elution capacity of the mobile phase is increased (Kazakevich et al. 2005; Kazakevich and Snow 2006). This can account for observed declines in retention factors with increasing organic modifier percentage, when higher percentages are used in the mobile phase. It is interesting to note that this decline is typically linear with methanol, and follows a more quadratic trend with acetonitrile and THF (Flieger 2007), consistent with the adsorption behavior of these solvents. Similarly, formation of an organic-rich surface layer creates a low dielectric environment, accounting for the apparently more efficient ion-pair formation when mobile phases containing acetonitrile are contrasted to those modified with methanol (Flieger and Świeboda 2008).

To summarize, understanding excess adsorption isotherm of the organic modifier is essential for optimizing mobile phase composition to achieve satisfactory separation. At lower content, increasing the mobile phase percentage of acetonitrile or THF will enhance charged analytes' retention via lowering the dielectric constant of the stationary phase surface, increasing the surface excess of the chaotropic agent, and favoring ion pairing in the stationary phase. At higher mobile phase content, increasing the organic modifier percentage will—linearly or quadratically—shorten analyte retention due to increased eluent strength.

1.4.3   EFFECT OF COLUMN TYPE

Choice of the chromatographic column affects the surface adsorption of chaotropic ions and the overall retention behavior of analytes. Kazakevich et al. studied the adsorption of various chaotropic anions in chromatographic systems defined by different column types, organic modifier types, and organic modifier percentage in the mobile phase (Kazakevich et al. 2005; Kazakevich and Snow 2006). Generally, increasing column hydrophobicity favors the adsorption of the chaotropic agent, in particular when highly chaotropic anions, such as PF_6^-, are considered. Accordingly, the surface excess of PF_6^- increases going from perfluorophenyl to alkylphenyl, and *n*-alkyl columns, with strongest interactions observed with graphite columns. There is also an intricate interplay between the column type and organic modifier present in these systems. As discussed earlier, acetonitrile is known to adsorb on the stationary phase surface forming multiple layers, and the thickness of this layer has also been shown to depend on column type, although surface excess maxima typically reach the same values.

These observations related to column-dependent adsorption behavior of chaotropic anions are also reflected on the retention behavior of charged analytes. Hashem and Jira analyzed a set of beta-adrenergic antagonists in several chromatographic systems (Hashem and Jira 2006). The authors observed that choice of the column significantly affects separation, but since mobile phase composition was varied concomitantly, conclusions as to the direct effects of stationary phase chemical nature are precluded. Vemić et al. (2015a) analyzed pramipexole and five of its impurities (four basic characters and one neutral—see Figure 1.9 later in this chapter), by varying column type, chaotropic salt, and its mobile phase concentration. The obtained data were modeled within the framework of the extended thermodynamic approach. Since the fitted coefficients have clear physical meaning, as highlighted earlier, the obtained results could be used to elucidate the extent to which hydrophobicity of the column affects the relevant equilibria; for this purpose c_2/c_3 ratio can be used. As can be seen in Figure 1.6, increasing hydrophobicity of the column strongly favors ion paring in the stationary phase. The relative magnitude of this effect was also found to be roughly proportional to the chaotropicity of the anion in question. Similarly, larger surface potentials are developed with more hydrophobic columns (Figure 1.4).

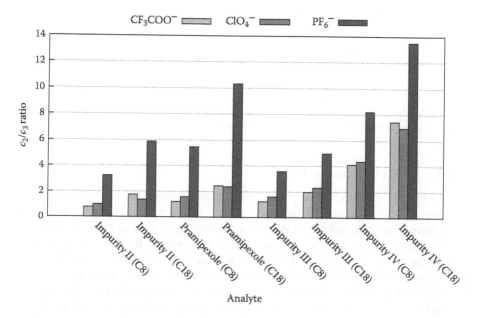

FIGURE 1.6 Variance in c_2/c_3 ratio as function of column type (C8 vs. C18) and choice of the chaotropic anion.

1.4.4 EFFECT OF MOBILE PHASE pH

The most readily appreciable effect of mobile phase pH on retention behavior in chaotropic chromatography is mediated by changes in ionization state of analytes. Lowering of pH shifts the ionization equilibrium of basic analytes toward protonated microspecies, favoring ion interactions, enhancing chaotropic effects, and resulting in longer retention times (LoBrutto et al. 2001b).

Čolović et al. (2015) first demonstrated that mobile phase pH can exhibit further more complex influence on retention in chaotropic chromatography. Namely, our group analyzed a set of diverse basic drugs in several RP-HPLC systems modified by addition of NaPF$_6$. The pH of the aqueous part of the mobile phase was varied from 2 to 4. Under each of the chosen chromatographic conditions, full protonation of the considered analytes was assured, meaning that any changes in retention factors would not be attributable to shifting in ionization equilibria of analytes. A fairly surprising behavior was observed, whereby fully protonated analytes experienced prolonged retention with increases in pH. In order to rationalize this finding, we investigated whether changes in pH affect adsorption of the chaotropic agent onto the stationary phase surface. Indeed, as illustrated in Figure 1.7, the surface excess of PF$_6^-$, estimated by the breakthrough method of frontal analysis, is consistently greater at higher pH values within the tested range. Since this was not something that could be explained by the chemical nature of the PF$_6^-$ anion itself, we leveraged the earlier discussed findings of Kazakevich et al. (Kazakevich et al. 2005; Kazakevich and Snow 2006) to test whether altered acetonitrile adsorption could also account for altered adsorption of the chaotropic agent. Increasing pH from

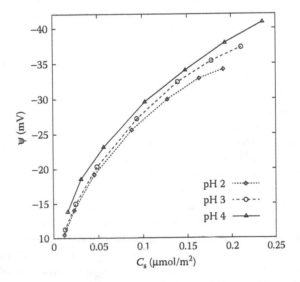

FIGURE 1.7 Differences in surface excess of PF$_6^-$ and the resulting surface potentials, occurring with changes in mobile phase pH. (From Čolović, J. et al., *J. Chromatogr. A*, 1425, 150–157, 2015.)

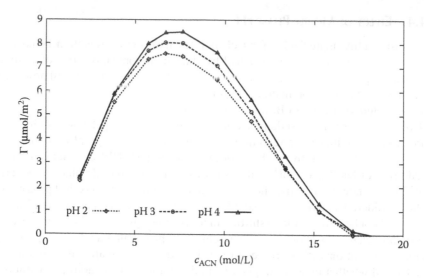

FIGURE 1.8 Excess adsorption isotherms of acetonitrile from acidified water at different pH values. (From Čolović, J. et al., *J. Chromatogr. A*, 1425, 150–157, 2015.)

2 to 4 was shown to significantly enhance acetonitrile surface excess (Figure 1.8), especially in the range of concentrations that were chosen for chromatographic analysis of the studied drugs.

How changes in pH modulate adsorption of acetronitrile is not entirely clear. One potential explanation is that residual silanol groups ionize at higher pH values, and that this affects the ordered nature of silanol-bound alkyl chains, which otherwise form high-affinity sites for acetonitrile binding (Kazakevich et al. 2001; Gritti and Guiochon 2005). If this was the case, however, it would also be reasonable to expect that electrostatic interactions would disfavor the adsorption of chaotropic anions, and conversely, favor the retention of oppositely charged analytes. The latter two effects seem to have much less significance for the overall retention behavior of charged solutes—than interactions with the acetonitrile surface layer. This became clear by studying pH-dependent changes in retention of the same set of analytes in the absence of chaotropic IIA. Singly charged analytes whose ionization state remains unchanged in the 2–4 pH range experience shortened retention times with increasing pH. This behavior is consistent with adsorption-partitioning model (Kazakevich et al. 2001), since increasing acetonitrile surface adsorption would create an increasingly organic-rich environment, disfavoring the partitioning of charged analytes. Conversely, analytes whose net positive charge increases with decreasing pH have longer retention times at pH 4 than at pH 2, following conventional RP-HPLC mechanism in the absence of chaotropic salt. These observations lend further support to the hypothesis that mobile phase pH effects, beyond those regarding analytes' ionization, are essentially mediated by changes in organic modifier's surface adsorption. Further experimental studies are needed to test this concept.

1.4.5 Effect of Mobile Phase Ionic Strength

Effects of changing ionic strength of the eluent have been studied in much detail when classical IIAs are concerned, and these effects can be rigorously modeled (Cecchi et al. 2004b). In principle, increasing the ionic strength leads to greater screening of all electrostatic interactions, and it effectively lowers the magnitude of the surface potential difference at any given distance from the stationary phase surface. The specifics of this phenomenon in chaotropic chromatography have been studied far less. Kazakevich et al. (2005) demonstrated that increasing ionic strength favors adsorption of PF_6^-. Most other studies tend to keep ionic strength constant while varying other factors.

1.5 MODELING THE RELATIONSHIP BETWEEN ANALYTES' STRUCTURE AND ITS RETENTION IN CHAOTROPIC CHROMATOGRAPHY

1.5.1 Quantitative Structure–Retention Relationship Modeling

As discussed in Section 1.3, although a thermodynamical treatment of retention data provides a means to accurately and rigorously model the influence of experimental conditions on the chromatographic behavior of individual analytes, this approach lacks predictive capabilities in which it is unable to account for how analytes' chemical structure affects their retention. Given the practical utility of such information, QSRR modeling has found widespread use in chromatographic studies. QSRR models essentially represent an extension of the concept of linear free energy relationships, whereby differences in analytes' chemical structure are assumed to lead to chemical potential differences, which are further translated into differences in physical observables, such as various retention parameters. This theoretical framework thus provides a basis for relating analytes' physicochemical properties to their chromatographic behavior in a quantitative manner. A detailed treatment of the QSRR paradigm and its applications can be found in several reviews (Héberger 2007; Kaliszan 2007; Put and Vander Heyden 2007). Here, we will briefly outline the key steps and caveats in building a QSRR model, with emphasis on issues of interest for the subsequent discussion on modeling retention behavior in chaotropic chromatography.

Independent variables of a typical QSRR model are descriptors, which numerically capture certain structural features of the considered molecules, with more than a thousand such indices proposed over the years (Todeschini and Consonni 2009). Some molecular descriptors encode information derived from the three-dimensional (3D) structure of the molecule, and such variables can, clearly, have utility in describing molecular phenomena that underlie retention mechanisms in chaotropic chromatography. However, a caveat that needs to be recognized is that 3D descriptors exhibit conformational dependence. Using single conformers in 3D QSRR analysis can potentially result in models of limited predictive capabilities (Hechinger et al. 2012), as typically several conformers are thermally accessible to a molecule in a chromatographic system, and choosing one *a priori* is often not possible.

This limitation can be overcome by using conformer ensembles and Boltzmann weighting as means to account for conformational diversity in the calculation of 3D descriptors (Oliveira et al. 2015).

The next step in establishing a QSRR model is choosing the most significant descriptors and applying a suitable regression or machine learning method to the matrix of descriptor values and retention parameters. Although many publications have addressed the question of optimal choice of method for variable selection and model building (Chen 2008; Riahi et al. 2009; Karanikolas et al. 2010; Fatemi and Elyasi 2011; Gupta et al. 2011; Goodarzi et al. 2012), there is no definitive rule in this respect. It is important to note that statistical performance of the model should not be the only determining factor in choosing a method and set of descriptors, as failing to interpret the resulting model can be seen as *a waste of the modeling effort* (Guha 2008). Multiple linear regression (MLR) is known to yield the most readily interpretable QSRR models, with performance that can often match that of more complex machine learning methods. Consequently, if a satisfactory model can be built using MLR, resorting to more elaborate techniques is usually not warranted. However, for modeling highly nonlinear retention phenomena, in complex chromatographic systems, methods such as artificial neural networks (ANN) and support vector machine (SVM) regression can provide the required performance, at some cost to model interpretability.

Although the uniformity of chromatographic retention data is an often praised quality that makes retention modeling a proving ground for various methodologies (Kaliszan 2007), it is also true that a typical chromatographic study is often not confined to the analysis of a given set of analytes under a single set of experimental conditions; mobile phase composition is often varied, either in terms of pH, organic modifier percentage, or the amount of added ion-interaction agent, as is of interest for the present review. When the experiments are designed in this manner, classical QSRR modeling cannot be used to analyze the data in its entirety—except through the generation of a series of locally applicable models. A more practical and general approach is to undertake the so-called *mixed modeling*, which involves incorporating mobile phase descriptors into the model as independent variables, in addition to structural descriptors. Numerous successful *mixed models* have been described (Tham and Agatonovic-Kustrin 2002; Ruggieri et al. 2005; D'Archivio et al. 2008; D'Archivio et al. 2014), demonstrating the successfulness and added utility of this approach.

Finally, validation is an integral part of developing a QSRR model (Tropsha et al. 2003; Dearden et al. 2009), as testing the model with data not used in training provides the relevant measures of its predictive capabilities and generalization potential. Similarly, defining the domain of model's applicability is an important step to undertake (Héberger 2007). For both QSRR and *mixed models*, it should be noted that the applicability domain cannot be defined solely in terms of chemical space, but that the specific experimental conditions, or range thereof, must also be taken into account.

1.5.2   QSRR Modeling in Chaotropic Chromatography

One of the earliest models that related analytes' structural characteristics to their retention times in RP-HPLC systems with chaotropic additives was proposed by

Li (2004b). This model extends the typical linear solvation free energy relationship (LSER) equation (reviewed in [Vitha and Carr 2006]) with two additional terms intended to account for the effects of analytes' ionization and ion pairing. The initial model, developed for fixed mobile phase composition, was later revised for applicability to linear gradient elution conditions, principally by simplification of the LSER term (Li and Rethwill 2004; Li 2004a). Both versions of the model include net charge as the sole descriptor of analytes' capability to participate in ion pairing. Although this can help separate neutral from charged solutes in terms of predicted retention times, the much larger and disparate standardized residuals that were observed for charged solutes clearly suggest that net charge provides a very coarse description of the ion-pairing affinity of different solutes. The entire data set was included in fitting of these models, which also involved removal of outliers, so no steps were undertaken to validate the predictions using external data. Finally, since a single ion-pairing term is used to account for all the ion interaction phenomena in this chromatographic system, it is questionable whether the resulting model can be considered as physically well founded.

Other authors analyzed the correlation between certain structural features of analytes and their retention parameters in ion-interaction chromatography, without actually developing predictive models. Such analyses tend to be useful, as they provide insights into the similar molecular determinants of retention that can guide subsequent variable selection, model development, and interpretation. In developing a modification of the extended thermodynamic approach applicable to modeling the retention of zwitterionic solutes, Cecchi et al. (2001c) described a good correlation between calculated molecular dipole moments and thermodynamic constants corresponding to stationary phase adsorption of these analytes. Essentially, the authors hypothesized that net charge neutrality of the zwitterion does not imply absence of electrostatic interactions with the charged stationary phase surface, since an effective fractional charge exists that facilitates interactions of the analyte with the surface charge of opposite sign (Cecchi et al. 2004a). Although this theory primarily helps rationalize the retention behavior of zwitterions in ion-interaction chromatography, it is interesting to note that charged solutes also have a quantifiable, permanent molecular dipole moment, potentially making the described findings of general interest in modeling effects of analytes' structure on its retention behavior.

Flieger performed a correlation analysis between retention factors of various beta-adrenergic antagonists, obtained under identical experimental conditions, and several calculated hydrophobicity indices of these structures (Flieger 2010). For the fairly congeneric set of analyzed compounds, a moderate to good correlation was observed, quantitatively supporting the empirical observations outlined in Section 1.4.1 regarding the increased chaotrope sensitivity of highly lipophilic analytes. Similar results were published for a set of phenothiazine and thioxanthene derivatives, where calculated effective partition coefficients at pH 7.4 ($\log D_{7.4}$) appeared to give best correlations to retention factors obtained in several RP-HPLC systems modified by either $NaPF_6$ or $NaClO_4$ (Flieger and Świeboda 2008). The authors, however, did not provide a rationale for using $\log D_{7.4}$ rather than $\log D_{2.8}$, which would better reflect the chosen experimental conditions. Instead, it was stipulated that good correlations

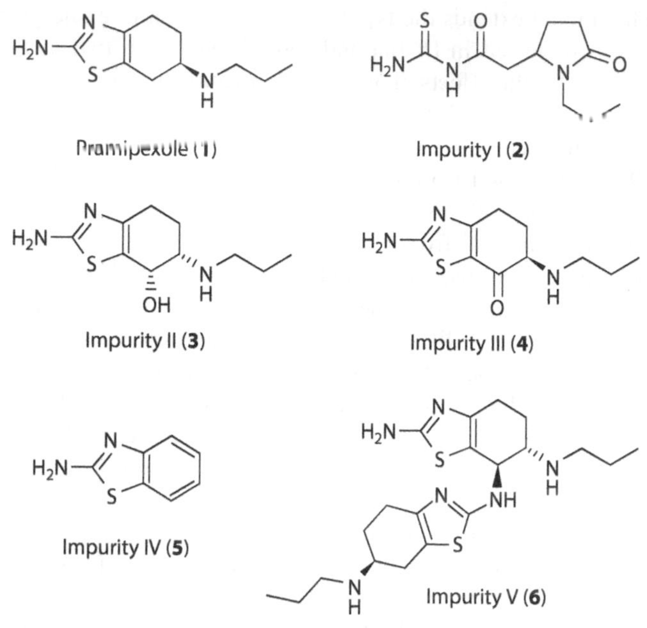

FIGURE 1.9 Structures of pramipexole and five of its impurities.

with this parameter suggest that there are similarities in partitioning of charged compounds in chromatographic and biological systems—whereby interactions with a counterion *neutralize* the compound thus enabling effective interactions with the stationary phase or lipophilic cellular compartments.

Considering hydrophobicity alone, however, is seldom sufficient to predict the elution order, even of structurally highly similar analytes. Our group began the study of the effects of analytes' structure on their retention behavior in chaotropic chromatography with the pharmaceutically relevant example of pramipexole (Figure 1.9, **1**) and its impurities (**2–6**) (Vemić et al. 2015a), previously discussed in Section 1.4.3. Although in absence of chaotropic salts, the elution order follows the predicted *n*-octanol–water distribution coefficient at pH 2.5 ($logD_{2.5}$), upon addition of the ClO_4^-, PF_6^-, or CF_3COO^-, this no longer holds. Neutral compound **2** experiences a slight retention decrease, explainable through surface exclusion phenomena (Cecchi et al. 2001a), whereas the remaining analytes' retention is prolonged, but not entirely consistent with their relative hydrophobicity. Specifically, **1** ($logD_{2.5} = -1.63$, $logP = 1.49$), **3** ($logD_{2.5} = -1.68$, $logP = 0.57$), and **4** ($logD_{2.5} = -1.32$, $logP = 0.94$) can be well separated but elute in the order: **3**, **1**, and lastly **4**. To gain further insights into the molecular determinants of such behavior, we correlated values of various descriptors with both retention factors and fitted parameters of the extended thermodynamic model (Table 1.1).

Although the importance of an aromaticity index (number of aromatic bonds, N_{AB}) could be readily interpreted in terms of retention mechanisms in RP-HPLC,

TABLE 1.1

Molecular Descriptors Best Correlated to Retention Factors and Fitted Parameters of the Extended Thermodynamic Model

	Retention Factor k		Parameter c_2		Parameter c_3	
Chaotrope	C8	C18	C8	C18	C8	C18
CH₃COONa	N_{AB}	N_{AB}	PNSA-1	PNSA-1	μ	β-POL
	$\beta_1 = 0.5274$	$\beta_1 = 1.3426$	$\beta_1 = 0.0018$	$\beta_1 = 0.0023$	$\beta_1 = 0.0077$	$\beta_1 = 0.0004$
	$\beta_0 = -1.8209$	$\beta_0 = -5.4759$	$\beta_0 = -0.0856$	$\beta_0 = -0.2018$	$\beta_0 = -0.0016$	$\beta_0 = 0.0657$
	$R^2 = 0.95$	$R^2 = 0.98$	$R^2 = 0.87$	$R^2 = 0.88$	$R^2 = 0.98$	$R^2 = 0.88$
	PNSA-1	PNSA-1				
	$\beta_1 = 0.0590$	$\beta_1 = 0.1435$				
	$\beta_0 = -5.6445$	$\beta_0 = -14.3470$				
	$R^2 = 0.90$	$R^2 = 0.84$				
NaClO₄	N_{AB}	N_{AB}	PNSA-1	PNSA-1	HASA-2/TMSA	FHACA
	$\beta_1 = 0.7378$	$\beta_1 = 1.5653$	$\beta_1 = 0.0014$	$\beta_1 = 0.0028$	$\beta_1 = 2.3226$	$\beta_1 = 1.1592$
	$\beta_0 = -2.7332$	$\beta_0 = -6.5086$	$\beta_0 = -0.1224$	$\beta_0 = -0.2472$	$\beta_0 = -0.0296$	$\beta_0 = -0.0008$
	$R^2 = 0.97$	$R^2 = 0.93$	$R^2 = 0.93$	$R^2 = 0.88$	$R^2 = 0.99$	$R^2 = 0.95$
	PNSA-1	PNSA-1				
	$\beta_1 = 0.0794$	$\beta_1 = 0.1647$				
	$\beta_0 = -7.6782$	$\beta_0 = -16.5160$				
	$R^2 = 0.85$	$R^2 = 0.77$				
NaPF₆	N_{AB}	N_{AB}	WNSA-1	WNSA-1	P''	P''
	$\beta_1 = 1.1818$	$\beta_1 = 4.1858$	$\beta_1 = 0.0130$	$\beta_1 = 0.0044$	$\beta_1 = 0.0861$	$\beta_1 = 0.0841$
	$\beta_0 = -4.8339$	$\beta_0 = -19.5520$	$\beta_0 = -0.5043$	$\beta_0 = -1.8990$	$\beta_0 = -0.0271$	$\beta_0 = -0.0348$
	$R^2 = 0.86$	$R^2 = 0.76$	$R^2 = 0.92$	$R^2 = 0.94$	$R^2 = 0.75$	$R^2 = 0.88$
	PNSA-1	PNSA-1				
	$\beta_1 = 0.1214$	$\beta_1 = 0.4189$				
	$\beta_0 = -12.0190$	$\beta_0 = -43.5690$				
	$R^2 = 0.69$	$R^2 = 0.57$				

Notes: β_1—regression parameter of the corresponding molecular descriptor; β_0—intercept; R^2—coefficient of determination; N_{AB}—number of aromatic bonds; *PNSA-1*—partially negatively charged surface area; *WNSA-1*—total molecular surface area (TMSA)-weighted partially negatively charged surface area (PNSA-1*TMSA/1000); μ—total dipole of the molecule; β-POL—first-order hyperpolarizability of the molecule; *HASA-2/TMSA*—area-weighted surface charge of hydrogen bonding acceptor atoms, divided by TMSA; *FHACA*—fractional area-weighted hydrogen bonding acceptor ability of the molecule; P''—polarity parameter ($Q_{max} - Q_{min}$) divided by the square of distance between the most positive (Q_{max}) and the most negative (Q_{min}) partial charge in the molecule.

the significance of some charged partial surface area descriptors was less than straightforward to rationalize. Specifically, the partially negatively charged surface area (*PNSA-1*) descriptor appeared to be an excellent predictor of retention factors and was found to be highly correlated to the c_2 parameter of the extended thermodynamic model, which corresponds to ion pairing in the stationary phase. What was surprising is the positive sign of the regression coefficients for *PNSA-1* that suggests that compounds with larger negative surface area are better retained and form ion pairs in the stationary phase more readily. Considering the surface excess of the negatively charged chaotrope, this finding appeared counterintuitive at first. However, if one considers the structure of the electric double layer, which forms on the surface of the stationary phase, it becomes clear that there is a possible mechanistic rationale; while electrostatic interactions of the positively charged segments of the molecule with the oppositely charged, adsorbed chaotrope provide the driving force for ion pairing, the electronegative segments of the analytes' are left to interact with the oppositely charged ions from the Stern layer. Thus, differences in *PNSA-1* can clearly account for differential retention behavior of structurally similar analytes, as illustrated in Figure 1.10 with the example of aforementioned compounds **1**, **3**, and **4**. These three compounds are all doubly protonated at pH 2.5, at which the experiments were performed, and have *PNSA-1* values of 99.24, 107.36, and 126.75, respectively. Whereas **4** is less lipophilic than **1**, its structure is such that partially negative surface area is well separated from the positive charge, and can interact

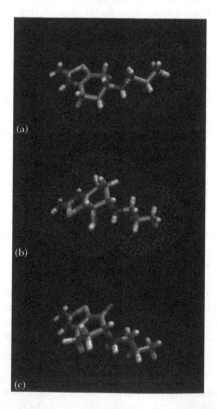

FIGURE 1.10 Molecular models of (a) compound **1**, (b) compound **3**, and (c) compound **4**. Dotted areas correspond to solvent accessible surface areas determined using a 1.4 Å probe, and colored according to partial atomic charges calculated with the AM1-BCC method, from geometries optimized *in vacuo*. Red areas represent partially negative, and blue areas represent partially positive atomic charges. (From Vemić, A. et al., *J. Chromatogr. A*, 1386, 39–46, 2015a.)

favorably with counterions from the Stern layer, thus accounting for longer retention. Conversely, the *syn* configuration of hydroxyl and amine substituents in **3** results in the partially negatively charged area projecting in the same direction as the dominant positive charge, which is less electrostatically favorable within the double layer and results in shortest retention times for this analyte. Consistent with this mechanistic interpretation, the long retention of compound **5** can also be explained through favorable separation of opposite charges and a large *PNSA-1* value of 146.24, resulting from efficient charge delocalization in an aromatic system.

It should be noted that the previous discussion clearly emphasizes the importance of considering charge distribution for QSRR modeling in chaotropic chromatography. Namely, it would seem that structural complementarity of analytes' electronic structure to that of the electric double layer plays a major role in determining its retention behavior. As this is not a structural characteristic accounted for explicitly in calculations of *PNSA-1*, this descriptor alone cannot be seen as sufficient in accounting for the effects of analytes' structure on retention. On the other hand, molecular dipole moments do depend on charge distribution, but our analysis did not suggest that this was statistically the most significant descriptor for quantifying the differences in c_2 values, at least not within the small set of analytes we initially considered. A composite index, or several descriptors, therefore, seems to be required in developing generally applicable models.

In our subsequent study, we attempted just this—to explore the possibility of establishing a more generally applicable QSRR model, while further investigating the molecular basis for differential retention behavior of charged solutes in chaotropic chromatography (Čolović et al. 2015). The modeling involved retention data obtained for 34 structurally diverse analytes in 36 chromatographic systems, defined by combinations of $NaPF_6$ concentration in the aqueous part of mobile phase, aqueous phase pH value, and acetonitrile content. In order to account for conformational effects in calculations of 3D descriptors, the aforementioned Boltzmann weighting approach was used. Model development was based on the *mixed model* paradigm, allowing for simultaneous modeling of data from all 36 chromatographic systems, and involved using SVM regression. The final model included four molecular descriptors and three mobile phase descriptors, and exhibited good performance both in training ($n = 938$, $R^2_{train} = 0.9577$, $RMSE_{train} = 0.1238$) and test sets ($n = 286$, $R^2_{test} = 0.9117$, $RMSE_{test} = 0.1806$). The four molecular descriptors included in the developed model are *ETA_Eta_B_RC* (branching index EtaB, with ring correction, relative to molecular size), *XlogP* (calculated octanol/water partition coefficient), *TDB9p* (3D topological distance-based autocorrelation—lag 9/weighted by polarizabilities), and *RDF45p* (radial distribution function—045/weighted by relative polarizabilities).

Of the four molecular descriptors listed, two can be seen to quantify structural parameters that can affect retention in any chromatographic system. Namely, *ETA_Eta_B_RC* is an extended topochemical index (Roy and Das 2011), which conveys information on how *branched* a given structure is. Greater degree of molecular branching can shorten retention times, as has been observed earlier (Tham and Agatonovic-Kustrin 2002) and rationalized in terms of steric effects and molecular bulkiness impeding effective interaction with the stationary phase. Similarly, *XlogP* accounts for the hydrophobicity of the structure, which has clear implications for retention in an RP-HPLC system.

The remaining two molecular descriptors in the model were seen as much more specific in accounting for specific effects in a chromatographic system with chaotropic additives. $TDB9p$ is a 3D topological distance-based autocorrelation descriptor (Klein et al. 2004) whose values are proportional to the sum, over all atom pairs 9 bonds apart, of the atomic polarizability products weighted by the respective geometrical distances. Similarly, $RDF45p$ quantifies the probability of finding two atoms at a distance of 4.5 Å in a given molecule, weighted by their relative polarizabilities. Both descriptors thus account for a specific spatial arrangement of polarizable atoms, albeit in two distinct distance bins. The composite nature of these indices makes them less intuitive to interpret than either $PNSA-1$ or the molecular dipole moment, but it is clear that both can be related to analytes' affinity for the stationary phase surface and complementarity with the electric double layer.

As can be seen from the preceding discussion, QSRR modeling has been used in ion-interaction chromatography in a very small number of studies, especially when these numbers are contrasted to modeling in conventional RP-HPLC. Although this is understandable given the inherent complexity of the system, foundations have been laid over the past few years that provide an incentive for new investigations in the field. What seems as a particularly attractive direction to explore is further analysis of structural factors that govern interactions with the electric double layer, ion pairing in the stationary, and mobile phases, respectively. Parameters of the extended thermodynamic model relate to the latter two phenomena directly, which is what makes them highly suited for further QSRR studies. The requirement for such investigations, however, is profiling the retention behavior of a larger number of diverse analytes in a way that enables the application of the extended thermodynamic model. This is experimentally fairly involved but can provide a rational foundation for development of physically well-grounded models that account for effects of analytes' structure on all the relevant equilibria in the system. It is important to emphasize that the extrathermodynamic nature of QSRR and mixed models do not imply that they should be developed with little regard to a good phenomenological description of the chromatographic system. Thus, while necessarily lacking the rigor of purely theoretical models, QSRR models can add major value to studies in chaotropic chromatography. Moreover, simple, reliable, and interpretable models of this kind can have significant practical utility in guiding selection and optimization of chromatographic conditions for separating a chosen set of analytes.

1.6 APPLICATIONS OF CHAOTROPIC CHROMATOGRAPHY IN DRUG ANALYSIS

In Sections 1.4 and 1.5, the influence of chaotropic agents and their interactions with other relevant chromatographic factors on the retention of basic analytes was extensively reviewed. For successful development of analytical methods intended for basic substances' quality control, the use of chaotropic agents is quite convenient because of their ability to manipulate analytes' retention without additional modifications of pH, column type, or organic modifier. Besides the increase of retention, chaotropic agents have one more very important feature—ability to improve chromatographic performance in terms of separation efficiency and peak shape of protonated basic drugs (Roberts et al. 2002; Pan et al. 2004; Wang and Carr 2007). A very well-known fact is that the separation of basic analytes on reversed-phase stationary phases is hampered by peak broadening and serious tailing due to the secondary interaction between basic solutes and residual silanol groups of column packing. The problem cannot be solved by the decrease of mobile phase pH value because it improves peak symmetry by suppression of silanol activity, while leading to unsatisfactory retention of the protonated analytes. However, the addition of chaotropic agent to the mobile phase was shown to be beneficial for overcoming this issue leading to the improvement in peak shape (Roberts et al. 2002). Generally, chaotropes enhance the loading capacity, which can be particularly useful for impurities profiling (Pan et al. 2004). Finally, chaotropic agents mimic the effect of classical amphiphilic ion-pair reagents, but due to their faster desorption from the stationary phase with increasing content of the organic modifier in the mobile phase, they can also be applied in gradient organic modifier elution methods.

In the review by Makarov et al. (2008) apart from the general overview of the processes in the chaotropic chromatography, the usefulness of chaotropic agents as mobile phase additives for optimization and fine tuning of complex separations for a wide variety of analytes, from small molecules to peptides and even chiral separations is emphasized. The practical advantages of chaotropic chromatography application in the analysis of selected basic analytes (LoBrutto et al. 2001b), alkaloides (Flieger 2006, 2007), β-blockers (Jones et al. 2002; Hashem and Jira 2006), tetracyclines and flumequine (Pilorz and Choma 2004), phenothiazine and thioxanthene derivatives (Flieger and Świeboda 2008), biogenic amines (Jolanta Flieger and Czajkowska-Żelazko 2011), and ionic liquids (Jolanta Flieger and Czajkowska-Żelazko 2012) were also elaborated. In method development mainly the traditional, quite tedious, one-factor-at-time approach was employed.

This approach is not only time consuming and economically inefficient but also provides insufficient and questionable data—if experiments are performed randomly, the result obtained will also be random (Lundstedt et al. 1998). Therefore, it is highly recommendable to apply systematical and efficient examination of all the factors that might have influence on a certain analytical problem. Experimental design is a very useful tool for screening of experimental variables in order to identify the influential ones, for the optimization of selected factor values, as well as for method robustness testing at the end of method development. In our earlier research, DoE methodology was applied to examine the influence of mobile phase composition, as well as the interactions of its components, on the chromatographic behavior of ropinirole and its two impurities (Vemić et al. 2014). Plan of experiments was defined by face-centered central composite design (CCD), a three-level factorial design suitable for collecting data aimed at elucidating the effects of selected factors, their interactions, and also for the definition of optimal conditions using grid point search method. We have also assessed the applicability of chaotropic agents in the separation of levodopa, carbidopa, entacapone, and their impurities, a complex mixture of analytes with different acid-base properties, employing gradient elution. Here, Box–Behnken experimental design was used for final tuning of the gradient program. This design is a three-level factorial design as well, suitable for the estimation of chosen factor effects, their interactions, and further optimization by desirability function (Vemić et al. 2013).

The benefits of using DoE methodology were recognized by regulatory authorities and included in contemporary guidance documents as a support in incorporating quality into pharmaceutical products. Pharmaceutical product development and analytical method development go hand in hand, because method provides critical data that support the understanding and control of pharmaceutical materials. Hence, we should consider analytical method as a process in which output is a reportable result, that is, the value that will be compared to the acceptance criterion. Therefore, methods are also required to meet quality demands postulated by the adopted regulatory documents (Food and Drug Administration 2003), ICH Q8 (ICH 2009), ICH Q9 (ICH 2005), and ICH Q10 (ICH 2008). The concept of quality by design (QbD) implemented in analytical method development improves the understanding and control of this process and hence its overall quality. As the implementation of analytical QbD (AQbD) concept in method development is novel and advantageous approach, we will explain it here in more detail. Figure 1.11 depicts the key phases involved in AQbD paradigm. Although the phases are shown in sequential manner, in practice some of the activities from one phase may overlap with another.

Case Study

The principle of chaotropic chromatography method development following AQbD principles is demonstrated on the separation of basic analytes: pramipexole and its five impurities (Figure 1.9), employing isocratic elution mode, while the particularities of method development when gradient elution is required are explained on the

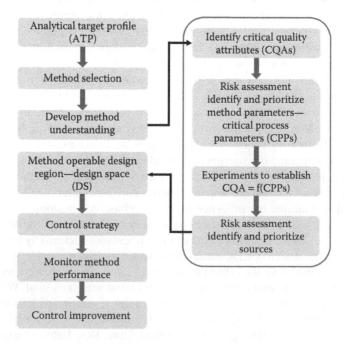

FIGURE 1.11   Key phases in AQbD approach.

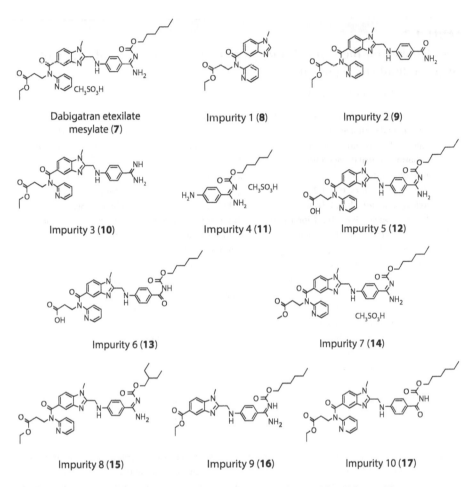

Dabigatran etexilate mesylate (**7**)

Impurity 1 (**8**)

Impurity 2 (**9**)

Impurity 3 (**10**)

Impurity 4 (**11**)

Impurity 5 (**12**)

Impurity 6 (**13**)

Impurity 7 (**14**)

Impurity 8 (**15**)

Impurity 9 (**16**)

Impurity 10 (**17**)

FIGURE 1.12 Structures of dabigatran etexilate mesylate and its 10 impurities.

separation of dabigatran etexilate mesylate, its degradation products, and process-related impurities (Figure 1.12).

The focus of the AQbD concept is the definition of the method operable design region or design space (DS) (ICH 2009). The foundation of AQbD is a predefined objective that stipulates the performance requirements for the analytical procedure called analytical target profile (ATP). The ATP should also comprise quality requirements, including the expected level of confidence for the reportable result that allows the correct conclusion to be drawn regarding the attributes of the method (Martin et al. 2016). In this stage, it is also necessary to set the critical quality attributes (CQA) that appropriately reflect method performance (Table 1.2).

Further on, risk assessment has to be applied in order to develop adequate method understanding. Various risk-assessment tools are suggested (ICH 2005) to identify potential variables that may need to be controlled to ensure procedure performance and to prioritize experimentation to eliminate or diminish areas of risk. The variables that could have an effect on method performance in chaotropic chromatographic

TABLE 1.2

Key Elements for the AQbD Paradigm Applied on Isocratic and Gradient HPLC Method Development

	Isocratic Elution Mode	Gradient Elution Mode
ATP	The efficient baseline separation and accurate determination of pramipexole and its five impurities from the pharmaceutical dosage form. *Probability*: $\pi \geq 95\%$ *Separation factors*: $\alpha \geq 1.2$ *Retention factors*: $\ln k_{II} > 0$ and $\ln k_V < 2.5$ *Recovery values*: 98%–102% for active ingredient, 70%–130% for related compounds. *Limit of detection*: not less than 0.05%	The efficient baseline separation and accurate determination of dabigatran etexilate mesylate and its 10 impurities from the pharmaceutical dosage form. *Probability*: $\pi \geq 95\%$ *Separation criterion*: $s \geq 0$ *Recovery values*: 98%–102% for active ingredient, 70%–130% for impurities with the specification limit of 0.5% (impurities 1–9), 90%–110% for impurities with the specification limit > 1% (impurity 10). *Limit of detection*: not less than 0.05%
CPP	A: concentration of the chaotropic agent (mM) B: acetonitrile content (%) C: column temperature (°C) D: XTerra C8 and XTerra C18 E: CH_3COONa and $NaClO_4$	x_1—content of the acetonitrile at the start of linear gradient ACN_{start} (%) x_2—content of the acetonitrile at the end of linear gradient ACN_{end} (%) x_3—gradient time t_G (min)
CQA	$\alpha_1 = k_I/k_{IV}$ or $\alpha_{11} = k_{IV}/k_I$ $\alpha_2 = k_V/k_{IV}$ or $\alpha_{22} = k_{IV}/k_V$ $\alpha_3 = k_I/k_V$ or $\alpha_{33} = k_V/k_I$ k_{II} k_V	$s_1 = t_{b\_imp2} - t_{e\_imp1}$ $s_2 = t_{b\_imp3} - t_{e\_imp2}$ $s_3 = t_{b\_imp5} - t_{e\_imp4}$ $s_4 = t_{b\_imp8} - t_{e\_imp7}$

Notes: ATP—analytical target profile; CPP—critical process parameters; CQA—critical quality attributes; α—selectivity factor for adjacent peak pairs; k_I, k_{II}, k_{IV}, k_V—retention factors of the impurities I, II, IV, and V, respectively; t_{b_imp}—retention time of impurity peak beginning; t_{e_imp}—retention time of impurity peak end.

systems were identified and analyzed to estimate risk associated with them. Ishikawa diagram was generated and control, noise, eXperimental (CNX) approach (Martin et al. 2016) was applied to classify all the variables (Figure 1.13).

Method development was conducted by experienced researchers with regular consultations within the analytical team. All the materials used in these two studies, except reference standards, as well as the HPLC system components were designated as noise factors. The risk associated with them is minimized by appropriate maintenance of equipment, by purchasing the chemicals of adequate purity and grade, and by appropriate storage and sampling. Appropriateness of the sample preparation procedure depends on sample properties. In this study, uncoated tablets containing pramipexole and capsules with dabigatran etexilate mesylate were analyzed. Therefore, quite conventional sample preparation procedure was employed and all the factors related to aforementioned were fixed, since they can be evaluated during method validation.

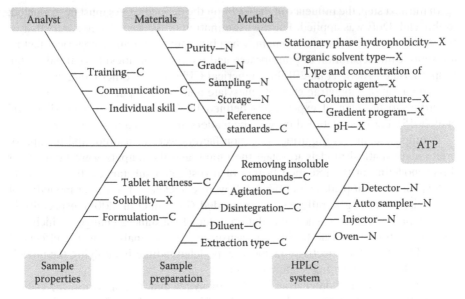

FIGURE 1.13 Risk assessment by CNX approach. C—Control, N—Noise factor, X—eXperimental parameters.

In order to determine factors significant for further investigation (CPPs) in case of separation of pramipexole and its impurities, several variables (Figure 1.13) were evaluated: vendor and hydrophobicity of the RP columns (Zorbax Extend C18, XBridge C18, Luna C18, XTerra C18, XTerra C8), nature (methanol, acetonitrile) and content (5%–20% v/v) of the organic modifier, type (CF_3COONa, $NaClO_4$, $NaPF_6$) and concentration (1–100 mM in the aqueous phase) of the chaotropic salts, pH (2–6) of the mobile phase, and column temperature (20°C–40°C). The pH value was fixed to 2.5 because small changes in the aforementioned region do not cause significant differences in the degree of analytes' protonation, thereby also simplifying the DoEs. Further modification of peaks' shape and retention can be achieved with chaotropic salts. Based on all these findings, three numerical CPPs (A, B, and C in Table 1.2) and two categorical CPPs (D and E in Table 1.2) were selected.

For separating dabigatran etexilate mesylate and its impurities, the following selection of CPPs was evaluated: the type of the chaotropic agent (CF_3COOH, $HClO_4$), concentration of the chaotropic agent (50 mM, 70 mM, 100 mM, 150 mM, and 200 mM), pH of the aqueous phase (2.0, 2.5, 3.0, and 3.5), content of the acetonitrile at the start of linear gradient—ACN_{start} (10%–30%), content of the acetonitrile at the end of linear gradient—ACN_{end} (48%–60%), and gradient time—t_G (8 min–15 min). As chaotropic agent CF_3COOH in the concentration of 150 mM was selected and further on fixed as C variable. Since the influence of pH variation proved to be insignificant once the protonation of the analytes is achieved, the pH of the aqueous phase was kept at 3.5 (C variable). The selected CPPs are given in Table 1.2 (x_1, x_2, and x_3).

In the next step, the influence of the CPPs on the defined CQAs must be evaluated; to this end, DoE was applied. For the examination of the knowledge space, that is, the investigated experimental domain, when both numerical and categorical factors are significant (separation of pramipexole and its five impurities) D-optimal design is appropriate (Vemić et al. 2015b). Numerical CPPs, on the other hand, critical for the separation of dabigatran etexilate mesylate and its 10 impurities, were analyzed according to the plan of experiments defined by Box–Behnken design (Pantović et al. 2015). For more detailed insight, the readers are referred to the cited papers.

Pramipexole and its impurities retention factors were measured, and the appropriate mathematical models to relate the inputs and the outputs were built. After direct modeling of the retention factors of investigated substances, the selectivity factors for adjacent peak pairs (Table 1.2) were calculated. The direct modeling of selectivity factor is generally not recommended (Lebrun et al. 2008), especially in the separation problems where the elution order of substances changes, which was the case in this study due to different sensitivity of the analytes to the effects of CPPs. Therefore, the selectivity factors are modeled indirectly applying the models obtained for corresponding retention factors.

In the separation of dabigatran etexilate mesylate and its impurities, the critical pairs in the separation appeared to be peaks of impurities **1** and **2**, impurities **2** and **3**, impurities **4** and **5**, as well as impurities **7** and **8**. The CQAs (Table 1.2) were indirectly modeled, that is they were calculated from the corresponding retention time of the end of first eluting peak and retention time of the beginning of second eluting peak (Dewé et al. 2004). The models obtained for corresponding retention times proved to have exceptional prediction abilities in the defined experimental domain, so the selected CQAs (s_1, s_2, s_3, and s_4) could be modeled indirectly.

Following the AQbD principles, besides the achievement of the desired quality level represented by CQAs, it is also important to ensure that these critical responses are within the predefined limits with an acceptable level of probability (confidence level). The robust optimization approach provides the assurance of method quality represented by such DS, which can be considered a *safe zone* where no significant changes of the CQA levels should be observed as a consequence of small, deliberate changes of method parameters. First, for each investigated combination of column and salt (four in total) in the separation of pramipexole and its impurities, the knowledge space was gridded by discretization of the numerical parameters: concentration of the chaotropic salt (mM) [35:3:65], acetonitrile content (%) [5:0.5:15], and column temperature (°C) [20:1:40]. Thus, the numbers of factor levels were 11, 21, and 21, respectively, giving 4851 investigated CPP combinations for each column-salt combination. On the other hand, for the separation of dabigatran etexilate mesylate and its impurities, experimental domain was divided in 21 levels for ACN_{start} × 21 levels for ACN_{end} × 9 levels for t_G = 3969 points to be analyzed. For each grid point, CQAs (Table 1.2) were calculated in MATLAB® 7.10.0 from the corresponding mathematical models.

Second, Monte Carlo simulations were used to compute the predictive probability for a given CQA to be greater than a desired threshold (Table 1.2). In the development of method for the separation of pramipexole and its impurities, we compared two approaches. First approach was to assess model uncertainty propagating the error equal to the variance of the residuals (Dispas et al. 2014; Boulanger 2016) when the error distribution was added to the mean predicted responses in order to obtain responses' distribution for each operating condition corresponding to a grid point. Second approach consisted in propagating the error in model coefficients' calculation (Rozet et al. 2013) when the error distribution equal to the calculated standard error was added to the model coefficient estimates in order to obtain responses' distribution for each operating condition corresponding to a grid point. In both cases, Monte Carlo simulation included 10,000 iterations. The regions of knowledge space having satisfactory values of all defined CQAs with the desired quality level (probability $\pi \geq 95\%$) were computed. Only the combinations with C18 column succeeded to give the corresponding DSs, which are presented in Figure 1.14. As it could be easily seen from the given figures, DSs obtained applying the second approach are more narrow suggesting more strict constraints when the errors originating from the model coefficients' calculation are taken into account. Such perspective encompasses more sources of variation originating from 10 model coefficients, thus leading to the greater reliability of the defined DS. If DS is defined in such manner, it can be formally presented with following equation:

$$DS = \left\{ x_0 \in \chi / E_\theta \left[P\left(CQAs \in \Lambda\right)/X = x_0, \theta \right] \geq \pi \right\} \qquad (1.10)$$

where:

χ is the knowledge space

Λ is the specification

π is the specified quality level

θ is the set of model parameters that include the uncertainty estimated by the statistical model

P and E correspond to the operands of probability and mathematical expectation, respectively

Once the DS is defined, working point is selected among all the points included in DS. The selection is completely based on the analyst appraisal because it could be the point with the best value of certain CQA or some point, which is suitable for experimental work due to lower consumption of resources. As a working point, we selected the point that requires less consumption of energy and materials: C18 column, 41 mM $NaClO_4$ in the aqueous part of the mobile phase, 8% ACN (v/v), and 25°C column temperature. The shape of the obtained region that simultaneously satisfies all defined criteria is irregular (Figure 1.14a), and this may lead to inappropriate separation if the slight changes in working parameters would occur. Therefore, a Monte Carlo simulation of the propagation of error from the factors to the responses was performed. Each operative parameter of the working point was replaced by randomly selected values from the uniform distribution of error equal to the variance that could occur in routine laboratory use of the method: ± 1 mM $NaClO_4$, $\pm 0.1\%$ ACN, and $\pm 1°C$. For the particular responses, probabilities of achieving the desired value of CQA were $\pi_{\ln(kII)} = 100\%$, $\pi_{\ln(kV)} = 99.5\%$, $\pi_{\alpha1} = 100\%$, $\pi_{\alpha33} = 100\%$, while the overall probability to simultaneously achieve the desired CQAs was $\pi = 99.5\%$, thus proving the robustness of the selected point.

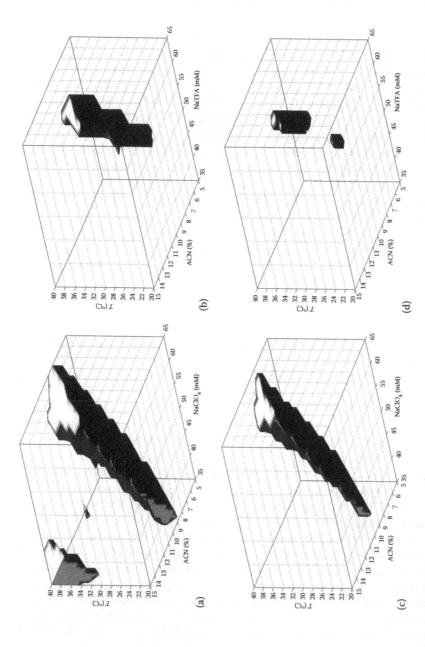

FIGURE 1.14 Three-dimensional representation of design space for predefined CQAs ($\ln k_{II} > 0$, $\ln k_V < 2.5$, and each $\alpha \geq 1.2$) achieved with probability $\pi \geq 95\%$: (a, b) propagating the error equal to the variance of the model residuals; (c, d) propagating the error originating from the model coefficients' calculation. (From Vemić, A. et al., *J. Pharm. Biomed. Anal.*, 102, 314–320, 2015b.)

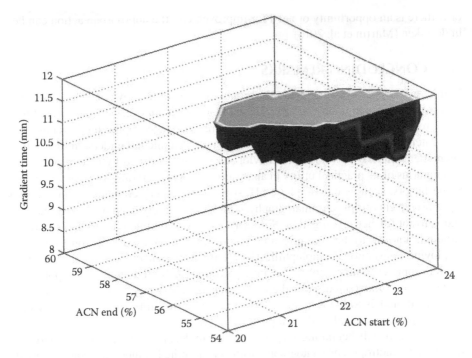

FIGURE 1.15 Three-dimensional representation of design space for predefined CQAs ($s_1 \geq 0$, $s_2 \geq 0$, $s_3 \geq 0$, and $s_4 \geq 0$) achieved with probability $\pi \geq 95\%$. (From Pantović, J. et al., *J. Pharm. Biomed. Anal.*, 111, 7–13, 2015.)

Based on the conclusions from the aforementioned research, DS of method intended for the separation of dabigatran etexilate mesylate and its impurities was acquired by the more strict approach (Equation 1.10). The uniform error distribution equal to the calculated standard error was added to the model coefficient estimates in order to obtain distribution of t_{b_imp2}, t_{e_imp1}, t_{b_imp3}, t_{e_imp2}, t_{b_imp5}, t_{e_imp4}, t_{b_imp8}, and t_{e_imp7} for each operating condition corresponding to a grid point (Rozet et al. 2013). Monte Carlo simulation included 5000 iterations in each of 3969 grid points. Through indirect modeling, the distributions of s_1, s_2, s_3, and s_4 for the given chromatographic conditions are also obtained. The region of knowledge space having satisfactory values of all defined CQAs with the desired quality level (probability $\pi \geq 95\%$) was computed (Figure 1.15). Point situated in the middle of the DS was chosen as the working point, thus providing the assurance of method robustness: 22% acetonitrile at the start of gradient program, 55.5% acetonitrile at the end of gradient program, and the gradient time of 11.5 minutes.

Finally, the developed methods were subjected to performance verification, that is, method validation—selectivity, linearity, range, precision, accuracy, limits of detection, and quantification were evaluated. As the results for all validation parameters were in accordance with regulatory requirements (Ahuja and Dong 2005), reliability of developed methods for use in routine pharmaceutical analysis was confirmed. For ongoing routine method, monitoring a set of SST parameters was defined. If, during routine monitoring of the procedure, data indicate that the procedure is out of control or if there is an opportunity or need for improvement, the appropriate action can be undertaken (Martin et al. 2016).

1.7 CONCLUDING REMARKS

In this review, we have attempted to summarize some of the key concepts related to chaotropic chromatography and provide the reader with an overview on some of the latest research efforts in the field. Study of this chromatographic mode remains a very interesting research subject, both in terms of fundamental understanding of the phenomena underlying retention behavior in systems modified by chaotropic salts, and in terms of its practical utility in analysis of complex mixtures of basic analytes. Some factors, such as mobile phase pH, seem to have very complex effects on the equilibria in chaotropic chromatography, which have largely been neglected thus far and warrant further investigation. Similarly, effects of mobile phase ionic strength, although known to be an influential factor in ion-interaction chromatography employing conventional amphiphilic IIAs, have not been thoroughly investigated in chaotropic chromatography. We anticipate that such studies have the potential to deepen our understanding of the underlying retention mechanisms, put existing theoretical models to the test, and also open novel possibilities in method development for pharmaceutical analytical purposes.

We are, also, only beginning to appreciate how molecular structure affects analytes' retention in chaotropic chromatography, with only a limited number of studies reported to date that have addressed this issue. Development of novel, physically well-grounded QSRR models is, thus, much needed. Such models can greatly expedite the selection of optimal chromatographic conditions for a given set of analytes, and these can provide useful insights into the nature of the dominant interactions between the analytes and components of the chromatographic system. Moving on, the prospect of studying chaotropic chromatography on an atomic scale, through the use of molecular simulations, provides an exciting incentive—for both chromatographers and computational chemists—to shed new light on an intriguing chromatographic system.

ACKNOWLEDGMENTS

The authors thank to the Ministry of Education, Science and Technological Development of the Republic of Serbia for supporting this investigation through the Projects 172052 and 172009.

REFERENCES

Ahuja, S. and M. Dong. 2005. *Handbook of Pharmaceutical Analysis by HPLC*. Amsterdam, the Netherlands: Elsevier.

Bartha, Á. and J. Ståhlberg. 1994. Electrostatic retention model of reversed-phase ion-pair chromatography. *Journal of Chromatography A*, 9th Danube Symposium on Chromatography, 668 (2): 255–284. doi:10.1016/0021-9673(94)80116-9.

Bidlingmeyer, B. A., S. N. Deming, W. P. Price Jr., B. Sachok, and M. Petrusek. 1979. Retention mechanism for reversed-phase ion-pair liquid chromatography. *Journal of Chromatography A* 186: 419–434. doi:10.1016/S0021-9673(00)95264-6.

Bidlingmeyer, B. A. 1980. Separation of ionic compounds by reversed-phase liquid chromatography an update of ion-pairing techniques. *Journal of Chromatographic Science* 18 (10): 525–539. doi:10.1093/chromsci/18.10.525.

Boulanger, B., E. Rozet, R. D. Marini, Ph. Hubert, P. Lebrun, T. Schofield, S. Rudaz, J. K. Mbinze, and B. Debrus. 2016. A new method for quality by design robust optimization in liquid chromatography. Accessed March 24. http://www.chromatographyonline.com/new-method-quality-design-robust-optimization-liquid-chromatography.

Bruzzoniti, M. C., E. Mentasti, G. Sacchero, and C. Sarzanini. 1996. Retention model for singly and doubly charged analytes in ion-interaction chromatography. *Journal of Chromatography A*, 19th International Symposium on Column Liquid Chromatography and Related Techniques, 728 (1–2): 55–65. doi:10.1016/0021-9673(95)01112-9.

Cantwell, F. F. 1984. Retention model for ion-pair chromatography based on double-layer ionic adsorption and exchange. *Journal of Pharmaceutical and Biomedical Analysis* 2 (2): 153–164. doi:10.1016/0731-7085(84)80066-7.

Cecchi, T. 2005. Use of lipophilic ion adsorption isotherms to determine the surface area and the monolayer capacity of a chromatographic packing, as well as the thermodynamic equilibrium constant for its adsorption. *Journal of Chromatography A* 1072 (2): 201–206. doi:10.1016/j.chroma.2005.03.022.

Cecchi, T. 2008. Ion pairing chromatography. *Critical Reviews in Analytical Chemistry* 38 (3): 161–213. doi:10.1080/10408340802038882.

Cecchi, T. 2009. *Ion-Pair Chromatography and Related Techniques.* Boca Raton, FL: CRC Press.

Cecchi, T. 2015. Theoretical models of ion pair chromatography: A close up of recent literature production. *Journal of Liquid Chromatography & Related Technologies* 38 (3): 404–414. doi:10.1080/10826076.2014.941267.

Cecchi, T., F. Pucciarelli, and P. Passamonti. 2004a. Ion-interaction chromatography of zwitterions. The fractional charge approach to model the influence of the mobile phase concentration of the ion-interaction reagent. *Analyst* 129 (11): 1037–1046. doi:10.1039/B404721D.

Cecchi, T., F. Pucciarelli, and P. Passamonti. 2004b. Extended thermodynamic approach to ion interaction chromatography. A mono- and bivariate strategy to model the influence of ionic strength. *Journal of Separation Science* 27 (15–16): 1323–1332. doi:10.1002/jssc.200401901.

Cecchi, T. and P. Passamonti. 2009. Retention mechanism for ion-pair chromatography with chaotropic reagents. *Journal of Chromatography A* 1216 (10): 1789–1797. doi:10.1016/j.chroma.2008.10.031.

Cecchi, T., F. Pucciarelli, and P. Passamonti. 2001a. Ion-interaction chromatography of neutral molecules. *Chromatographia* 53 (1–2): 27–34. doi:10.1007/BF02492423.

Cecchi, T., F. Pucciarelli, and P. Passamonti. 2001b. Extended thermodynamic approach to ion interaction chromatography. *Analytical Chemistry* 73 (11): 2632–2639. doi:10.1021/ac001341y.

Cecchi, T., F. Pucciarelli, P. Passamonti, and P. Cecchi. 2001c. The dipole approach in ion-interaction chromatography of zwitterions. *Chromatographia* 54 (1–2): 38–44. doi:10.1007/BF02491830.

Chen, H. F. 2008. Quantitative predictions of gas chromatography retention indexes with support vector machines, radial basis neural networks and multiple linear regression. *Analytica Chimica Acta* 609 (1): 24–36. doi:10.1016/j.aca.2008.01.003.

Chen, J. G., S. G. Weber, L. L. Glavina, and F. F. Cantwell. 1993. Electrical double-layer models of ion-modified (ion-Pair) reversed-phase liquid chromatography. *Journal of Chromatography A* 656 (1–2): 549–576. doi:10.1016/0021-9673(93)80819-T.

Collins, K. D. and M. W. Washabaugh. 1985. The Hofmeister effect and the behaviour of water at interfaces. *Quarterly Reviews of Biophysics* 18 (4): 323–422. doi:10.1017/S0033583500005369.

Čolović, J., M. Kalinić, A. Vemić, S. Erić, and A. Malenović. 2015. Investigation into the phenomena affecting the retention behavior of basic analytes in chaotropic chromatography: Joint effects of the most relevant chromatographic factors and analytes' molecular properties. *Journal of Chromatography A* 1425: 150–157. doi:10.1016/j. chroma.2015.11.027.

Dai, J. and P. W. Carr. 2005. Role of ion pairing in anionic additive effects on the separation of cationic drugs in reversed-phase liquid chromatography. *Journal of Chromatography A* 1072 (2): 169–184. doi:10.1016/j.chroma.2005.03.005.

Dai, J., S. D. Mendonsa, M. T. Bowser, C. A. Lucy, and P. W. Carr. 2005. Effect of anionic additive type on ion pair formation constants of basic pharmaceuticals. *Journal of Chromatography A* 1069 (2): 225–234. doi:10.1016/j.chroma.2005.02.030.

D'Archivio, A. A., M. A. Maggi, P. Mazzeo, and F. Ruggieri. 2008. Quantitative structure–retention relationships of pesticides in reversed-phase high-performance liquid chromatography based on WHIM and GETAWAY molecular descriptors. *Analytica Chimica Acta* 628 (2): 162–172. doi:10.1016/j.aca.2008.09.018.

D'Archivio, A. A., M. A. Maggi, and F. Ruggieri. 2014. Prediction of the retention of s-triazines in reversed-phase high-performance liquid chromatography under linear gradient-elution conditions. *Journal of Separation Science* 37 (15): 1930–1936. doi:10.1002/jssc.201400346.

Dearden, J. C., M. T. D. Cronin, and K. L. E. Kaiser. 2009. How not to develop a quantitative structure-activity or structure-property relationship (QSAR/QSPR). *SAR and QSAR in Environmental Research* 20 (3–4): 241–266. doi:10.1080/10629360902949567.

Deelder, R. S. and J. H. M. van den Berg. 1981. Study on the retention of amines in reversed-phase ion-pair chromatography on bonded phases. *Journal of Chromatography A* 218: 327–339. doi:10.1016/S0021-9673(00)82063-4.

Dewé, W., R. D. Marini, P. Chiap, Ph. Hubert, J. Crommen, and B. Boulanger. 2004. Development of response models for optimising HPLC methods. *Chemometrics and Intelligent Laboratory Systems* 74 (2): 263–268. doi:10.1016/j.chemolab.2004.04.016.

Diamond, R. M. 1963. The aqueous solution behavior of large univalent ions. A new type of ion-pairing. *The Journal of Physical Chemistry* 67 (12): 2513–2517. doi:10.1021/j100806a002.

Dispas, A., P. Lebrun, B. Andri, E. Rozet, and P. Hubert. 2014. Robust method optimization strategy—a useful tool for method transfer: The case of SFC. *Journal of Pharmaceutical and Biomedical Analysis* 88: 519–524. doi:10.1016/j.jpba.2013.09.030.

Eksborg, S., P. O. Lagerström, R. Modin, and G. Schill. 1973. Ion-pair chromatography of organic compounds. *Journal of Chromatography A* 83: 99–110. doi:10.1016/S0021-9673(00)97031-6.

Fatemi, M. H. and M. Elyasi. 2011. Prediction of gas chromatographic retention indices of some amino acids and carboxylic acids from their structural descriptors. *Journal of Separation Science* 34 (22): 3216–3220. doi:10.1002/jssc.201100544.

Fini, A., G. Fazio, M. Gonzalez-Rodriguez, C. Cavallari, N. Passerini, and L. Rodriguez. 1999. Formation of ion-pairs in aqueous solutions of diclofenac salts. *International Journal of Pharmaceutics* 187 (2): 163–173. doi:10.1016/S0378-5173(99)00180-5.

Flieger, J. 2006. The effect of chaotropic mobile phase additives on the separation of selected alkaloids in reversed-phase high-performance liquid chromatography. *Journal of Chromatography A* 1113 (1–2): 37–44. doi:10.1016/j.chroma.2006.01.090.

Flieger, J. 2007. Effect of mobile phase composition on the retention of selected alkaloids in reversed-phase liquid chromatography with chaotropic salts. *Journal of Chromatography A* 1175 (2): 207–216. doi:10.1016/j.chroma.2007.10.036.

Flieger, J. 2010. Application of perfluorinated acids as ion-pairing reagents for reversed-phase chromatography and retention-hydrophobicity relationships studies of selected β-blockers. *Journal of Chromatography A* 1217 (4): 540–549. doi:10.1016/j. chroma.2009.11.083.

Flieger, J. and A. Czajkowska-Żelazko. 2011. Comparison of chaotropic salt and ionic liquid as mobile phase additives in reversed-phase high-performance liquid chromatography of biogenic amines. *Journal of Separation Science* 34 (7): 733–739. doi:10.1002/jssc.201000797.

Flieger, J. and A. Czajkowska-Żelazko. 2012. Identification of ionic liquid components by RP-HPLC with diode array detector using chaotropic effect and perturbation technique. *Journal of Separation Science* 35 (2): 248–255. doi:10.1002/jssc.201100751.

Flieger, J. and R. Świeboda. 2008. Application of chaotropic effect in reversed-phase liquid chromatography of structurally related phenothiazine and thioxanthene derivatives. *Journal of Chromatography A* 1192 (2): 218–224. doi:10.1016/j.chroma.2008.02.117.

Food and Drug Administration. 2003. Pharmaceutical cGMPS for the 21st Century—a risk-based approach: Second progress report and implementation plan. http://www.fda.gov/Drugs/DevelopmentApprovalProcess/Manufacturing/QuestionsandAnswersonCurrentGoodManufacturingPracticescGMPforDrugs/UCM071836.

Ghaemi, Y. and R. A. Wall. 1979. Hydrophobic chromatography with dynamically coated stationary phases. *Journal of Chromatography A* 174 (1): 51–59. doi:10.1016/S0021-9673(00)87036-3.

Goodarzi, M., R. Jensen, and Y. Vander Heyden. 2012. QSRR modeling for diverse drugs using different feature selection methods coupled with linear and nonlinear regressions. *Journal of Chromatography B* 910: 84–94. doi:10.1016/j.jchromb.2012.01.012.

Gritti, F. and G. Guiochon. 2005. Adsorption mechanism in RPLC. Effect of the nature of the organic modifier. *Analytical Chemistry* 77 (13): 4257–4272. doi:10.1021/ac0580058.

Guha, R. 2008. On the interpretation and interpretability of quantitative structure–activity relationship models. *Journal of Computer-Aided Molecular Design* 22 (12): 8570–8571. doi:10.1007/s10822-008-9240-5.

Gupta, V. K., H. Khani, B. Ahmadi-Roudi, S. Mirakhorli, E. Fereyduni, and S. Agarwal. 2011. Prediction of capillary gas chromatographic retention times of fatty acid methyl esters in human blood using MLR, PLS and back-propagation artificial neural networks. *Talanta* 83 (3): 1014–1022. doi:10.1016/j.talanta.2010.11.017.

Hashem, H. and T. Jira. 2006. Effect of chaotropic mobile phase additives on retention behaviour of beta-blockers on various reversed-phase high-performance liquid chromatography columns. *Journal of Chromatography A* 1133 (1–2): 69–75. doi:10.1016/j.chroma.2006.07.074.

Héberger, K. 2007. Quantitative structure–(chromatographic) retention relationships. *Journal of Chromatography A* 1158 (1–2): 273–305. doi:10.1016/j.chroma.2007.03.108.

Hechinger, M., K. Leonhard, and W. Marquardt. 2012. What is wrong with quantitative structure–property relations models based on three-dimensional descriptors? *Journal of Chemical Information and Modeling* 52 (8): 1984–1993. doi:10.1021/ci300246m.

Horvath, C., W. Melander, I. Molnar, and P. Molnar. 1977. Enhancement of retention by ion-pair formation in liquid chromatography with nonpolar stationary phases. *Analytical Chemistry* 49 (14): 2295–2305. doi:10.1021/ac50022a048.

Hung, C. T. and R. B. Taylor. 1980. Mechanism of retention of acidic solutes by octadecyl silica using quaternary ammonium pairing ions as ion exchangers. *Journal of Chromatography A* 202 (3): 333–345. doi:10.1016/S0021-9673(00)91817-X.

ICH. 2005. ICH topic Q9: Quality risk management. http://www.ich.org/products/guidelines/quality/article/quality-guidelines.html.

ICH. 2008. ICH topic Q10: Pharmaceutical quality system.

ICH. 2009. ICH topic Q8 (R2): Pharmaceutical development. http://www.ich.org/products/guidelines/quality/article/quality-guidelines.html.

Jones, A., R. LoBrutto, and Y. Kazakevich. 2002. Effect of the counter-anion type and concentration on the liquid chromatography retention of β-blockers. *Journal of Chromatography A* 964 (1–2): 179–187. doi:10.1016/S0021-9673(02)00448-X.

Kaliszan, R. 2007. QSRR: Quantitative structure-(chromatographic) retention relationships. *Chemical Reviews* 107 (7): 3212–3246. doi:10.1021/cr068412z.

Karanikolas, B. D., M. L. Figueiredo, and L. Wu. 2010. Comprehensive evaluation of the role of EZH2 in the growth, invasion, and aggression of a panel of prostate cancer cell lines. *The Prostate* 70 (6): 675–688. doi:10.1002/pros.21112.

Kazakevich, I. I., and N. H. Snow. 2006. Adsorption behavior of hexafluorophosphate on selected bonded phases. *Journal of Chromatography A*, 29th International Symposium on High Performance Liquid Phase Separations and Related Techniques. Part I, 1119 (1–2): 43–50. doi:10.1016/j.chroma.2006.02.094.

Kazakevich, Y. V., R. LoBrutto, F. Chan, and T. Patel. 2001. Interpretation of the excess adsorption isotherms of organic eluent components on the surface of reversed-phase adsorbents: Effect on the analyte retention. *Journal of Chromatography A*, International Symposium on High Performance Liquid Phase Separations and Related Techniques. Part I, 913 (1–2): 75–87. doi:10.1016/S0021-9673(00)01239-5.

Kazakevich, Y. V., R. LoBrutto, and R. Vivilecchia. 2005. Reversed-phase high-performance liquid chromatography behavior of chaotropic counteranions. *Journal of Chromatography A* 1064 (1): 9–18. doi:10.1016/j.chroma.2004.11.104.

Kissinger, P. T. 1977. Comments on reverse-phase ion-pair partition chromatography. *Analytical Chemistry* 49 (6): 883. doi:10.1021/ac50014a054.

Klein, C. T., D. Kaiser, and G. Ecker. 2004. Topological distance based 3D descriptors for use in QSAR and diversity analysis. *Journal of Chemical Information and Computer Sciences* 44 (1): 200–209. doi:10.1021/ci0256236.

Knox, J. H. and R. A. Hartwick. 1981. Mechanism of ion-pair liquid chromatography of amines, neutrals, zwitterions and zcids using anionic hetaerons. *Journal of Chromatography A* 204: 3–21. doi:10.1016/S0021-9673(00)81633-7.

Knox, J. H. and G. R. Laird. 1976. Soap chromatography—a new high-performance liquid chromatographic technique for separation of ionizable materials. *Journal of Chromatography A* 122: 17–34. doi:10.1016/S0021-9673(00)82234-7.

Kraak, J. C., K. M. Jonker, and J. F. K. Huber. 1977. Solvent-generated ion-exchange systems with anionic surfactants for rapid separations of amino acids. *Journal of Chromatography A* 142: 671–688. doi:10.1016/S0021-9673(01)92076-X.

Lebrun, P., B. Govaerts, B. Debrus, A. Ceccato, G. Caliaro, P. Hubert, and B. Boulanger. 2008. Development of a new predictive modelling technique to find with confidence equivalence zone and design space of chromatographic analytical methods. *Chemometrics and Intelligent Laboratory Systems*, Selected papers presented at the Chemometrics Congress "CHIMIOMETRIE 2006" Paris, France, 30 November–1 December 2006, 91 (1): 4–16. doi:10.1016/j.chemolab.2007.05.010.

Li, J. 2004a. Prediction of internal standards in reversed-phase liquid chromatography. *Chromatographia* 60 (1–2): 63–71. doi:10.1365/s10337-004-0327-4.

Li, J. 2004b. Prediction of internal standards in reversed-phase liquid chromatography: IV. correlation and prediction of retention in reversed-phase ion-pair chromatography based on linear solvation energy relationships. *Analytica Chimica Acta* 522 (1): 113–126. doi:10.1016/j.aca.2004.06.043.

Li, J. and P. A. Rethwill. 2004. Systematic selection of internal standard with similar chemical and UV properties to drug to be quantified in serum samples. *Chromatographia* 60 (7–8): 391–397. doi:10.1365/s10337-004-0392-8.

Liu, H. and F. F. Cantwell. 1991. Electrical double-layer model for ion-pair chromatographic retention on octadecylsilyl bonded phases. *Analytical Chemistry* 63 (18): 2032–2037. doi:10.1021/ac00018a026.

LoBrutto, R., A. Jones, and Y. V. Kazakevich. 2001a. Effect of counter-anion concentration on retention in high-performance liquid chromatography of protonated basic analytes. *Journal of Chromatography A*, International Symposium on High Performance Liquid Phase Separations and Related Techniques. Part I, 913 (1–2): 189–196. doi:10.1016/S0021-9673(00)01031-1.

LoBrutto, R., A. Jones, Y. V. Kazakevich, and H. M. McNair. 2001b. Effect of the eluent pH and acidic modifiers in high-performance liquid chromatography retention of basic analytes. *Journal of Chromatography A*, International Symposium on High Performance Liquid Phase Separations and Related Techniques. Part I, 913 (1–2): 173–187. doi:10.1016/S0021-9673(00)01012-8.

Lo Nostro, P. and B. W. Ninham. 2012. Hofmeister phenomena: An update on ion specificity in biology. *Chemical Reviews* 112 (4): 2286–2322. doi:10.1021/cr200271j.

Lundstedt, T., E. Seifert, L. Abramo, B. Thelin, Å. Nyström, J. Pettersen, and R. Bergman. 1998. Experimental design and optimization. *Chemometrics and Intelligent Laboratory Systems* 42 (1–2): 3–40. doi:10.1016/S0169-7439(98)00065-3.

Makarov, A., R. LoBrutto, and Y. Kazakevich. 2008. Liophilic mobile phase additives in reversed phase HPLC. *Journal of Liquid Chromatography & Related Technologies* 31 (11–12): 1533–1567. doi:10.1080/10826070802125918.

Manallack, D. T. 2009. The acid–base profile of a contemporary set of drugs: Implications for drug discovery. *SAR and QSAR in Environmental Research* 20 (7–8): 611–655. doi:10.1080/10629360903438313.

Marcus, Y. 2009. Effect of ions on the structure of water: Structure making and breaking. *Chemical Reviews* 109 (3): 1346–1370. doi:10.1021/cr8003828.

Marcus, Y. and G. Hefter. 2006. Ion pairing. *Chemical Reviews* 106 (11): 4585–4621. doi:10.1021/cr040087x.

Martin, G. P., K. L. Barnett, C. Burgess, P. D. Curry, J. Ermer, G. S. Gratz, J. P Hammond. et al. 2016. Lifecycle management of analytical procedures: Method development, procedure performance qualification, and procedure performance verification. http://www.usp.org/sites/default/files/usp_pdf/EN/USPNF/revisions/lifecycle_pdf.pdf.

Oliveira, T. B., L. Gobbo-Neto, T. J. Schmidt, and F. B. Da Costa. 2015. Study of chromatographic retention of natural terpenoids by chemoinformatic tools. *Journal of Chemical Information and Modeling* 55 (1): 26–38. doi:10.1021/ci500581q.

Pan, L., R. LoBrutto, Y. V. Kazakevich, and R. Thompson. 2004. Influence of inorganic mobile phase additives on the retention, efficiency and peak symmetry of protonated basic compounds in reversed-phase liquid chromatography. *Journal of Chromatography A* 1049 (1–2): 63–73. doi:10.1016/j.chroma.2004.07.019.

Pantović, J., A. Malenović, A. Vemić, N. Kostić, and M. Medenica. 2015. Development of liquid chromatographic method for the analysis of dabigatran etexilate mesilate and its ten impurities supported by quality-by-design methodology. *Journal of Pharmaceutical and Biomedical Analysis* 111: 7–13. doi:10.1016/j.jpba.2015.03.009.

Parsons, D. F., M. Boström, P. L. Nostro, and B. W. Ninham. 2011. Hofmeister effects: Interplay of hydration, nonelectrostatic potentials, and ion size. *Physical Chemistry Chemical Physics* 13 (27): 12352–12367. doi:10.1039/C1CP20538B.

Pilorz, K. and I. Choma. 2004. Isocratic reversed-phase high-performance liquid chromatographic separation of tetracyclines and flumequine controlled by a chaotropic effect. *Journal of Chromatography A*, 27th International Symposium on High-Performance Liquid-Phase Separations and Related Techniques. Part II, 1031 (1–2): 303–5. doi:10.1016/j.chroma.2003.12.024.

Put, R. and Y. V. Heyden. 2007. Review on modelling aspects in reversed-phase liquid chromatographic quantitative structure–retention relationships. *Analytica Chimica Acta* 602 (2): 164–172. doi:10.1016/j.aca.2007.09.014.

Riahi, S., E. Pourbasheer, M. R. Ganjali, and P. Norouzi. 2009. Investigation of different linear and nonlinear chemometric methods for modeling of retention index of essential oil components. Concerns to support vector machine. *Journal of Hazardous Materials* 166 (2–3): 853–859. doi:10.1016/j.jhazmat.2008.11.097.

Roberts, J. M., A. R. Diaz, D. T. Fortin, J. M. Friedle, and S. D. Piper. 2002. Influence of the Hofmeister series on the retention of amines in reversed-phase liquid chromatography. *Analytical Chemistry* 74 (19): 4927–4932. doi:10.1021/ac0256944.

Roy, K. and R. N. Das. 2011. On some novel extended topochemical atom (ETA) parameters for effective encoding of chemical information and modelling of fundamental physicochemical properties. *SAR and QSAR in Environmental Research* 22 (5–6): 451–472. doi:10.1080/1062936X.2011.569900.

Rozet, E., P. Lebrun, P. Hubert, B. Debrus, and B. Boulanger. 2013. Design spaces for analytical methods. *TrAC Trends in Analytical Chemistry* 42: 157–167. doi:10.1016/j.trac.2012.09.007.

Ruggieri, F., A. A. D'Archivio, G. Carlucci, and P. Mazzeo. 2005. Application of artificial neural networks for prediction of retention factors of triazine herbicides in reversed-phase liquid chromatography. *Journal of Chromatography A* 1076 (1–2): 163–169. doi:10.1016/j.chroma.2005.04.038.

Sarzanini, C., M. C. Bruzzoniti, G. Sacchero, and E. Mentasti. 1996. Retention model for anionic, neutral, and cationic analytes in reversed-phase ion interaction chromatography. *Analytical Chemistry* 68 (24): 4494–4500. doi:10.1021/ac960536w.

Ståhlberg, J. 1986. The Gouy–Chapman theory in combination with a modified langmuir isotherm as a theoretical model for ion-pair chromatography. *Journal of Chromatography A* 356: 231–245. doi:10.1016/S0021-9673(00)91485-7.

Ståhlberg, J. 1999. Retention models for ions in chromatography. *Journal of Chromatography A* 855 (1): 3–55. doi:10.1016/S0021-9673(99)00176-4.

Ståhlberg, J. 2010. Ion-pair chromatography and related techniques. *Journal of the American Chemical Society* 132 (5): 1732. doi:10.1021/ja1000455.

Steiner, S. A., D. M. Watson, and J. S. Fritz. 2005. Ion association with alkylammonium cations for separation of anions by capillary electrophoresis. *Journal of Chromatography A*, 17th International Ion Chromatography Symposium 17th International Ion Chromatography Symposium, 1085 (1): 170–175. doi:10.1016/j.chroma.2005.01.070.

Takayanagi, T., E. Wada, and S. Motomizu. 1997. Electrophoretic mobility study of ion association between aromaticanions and quaternary ammonium ions in aqueous solution. *Analyst* 122 (1): 57–62. doi:10.1039/A605047F.

Tham, S. Y. and S. Agatonovic-Kustrin. 2002. Application of the artificial neural network in quantitative structure–gradient elution retention relationship of phenylthiocarbamyl amino acids derivatives. *Journal of Pharmaceutical and Biomedical Analysis* 28 (3–4): 581–590. doi:10.1016/S0731-7085(01)00690-2.

Tilly Melin, A., M. Liungcrantz, and G. Schill. 1979. Reversed-phase on-Pair chromatography with an adsorbing stationary phase and a hydrophobic quaternary ammonium ion in the mobile phase. *Journal of Chromatography A* 185: 225–239. doi:10.1016/S0021-9673(00)85606-X.

Todeschini, R. and V. Consonni. 2009. *Molecular Descriptors for Chemoinformatics.* 2nd ed. Vol. 41. Methods and Principles in Medicinal Chemistry. Weinheim: John Wiley & Sons.

Tomlinson, E., T. M. Jefferies, and C. M. Riley. 1978. Ion-pair high-performance liquid chromatography. *Journal of Chromatography A* 159 (3): 315–358. doi:10.1016/S0021-9673(00)91700-X.

Tropsha, A., P. Gramatica, and V. K. Gombar. 2003. The importance of being earnest: Validation is the absolute essential for successful application and interpretation of QSPR models. *QSAR & Combinatorial Science* 22 (1): 69–77. doi:10.1002/qsar.200390007.

Vemić, A., M. Kalinić, S. Erić, A. Malenović, and M. Medenica. 2015a. The influence of salt chaotropicity, column hydrophobicity and analytes' molecular properties on the retention of pramipexole and its impurities. *Journal of Chromatography A* 1386: 39–46. doi:10.1016/j.chroma.2015.01.078.

Vemić, A., A. Malenović, and M. Medenica. 2014. The influence of inorganic salts with chaotropic properties on the chromatographic behavior of ropinirole and its two impurities. *Talanta* 123: 122–127. doi:10.1016/j.talanta.2014.02.006.

Vemić, A., T. Rakić, A. Malenović, and M. Medenica. 2015b. Chaotropic salts in liquid chromatographic method development for the determination of pramipexole and its impurities following quality-by-design principles. *Journal of Pharmaceutical and Biomedical Analysis* 102: 314–320. doi:10.1016/j.jpba.2014.09.031.

Vemić, A., B. J. Stojanović, I. Stamenković, and A. Malenović. 2013. Chaotropic agents in liquid chromatographic method development for the simultaneous analysis of levodopa, carbidopa, entacapone and their impurities. *Journal of Pharmaceutical and Biomedical Analysis* 77: 9–15. doi:10.1016/j.jpba.2013.01.007.

Vitha, M. and P. W. Carr. 2006. The chemical interpretation and practice of linear solvation energy relationships in chromatography. *Journal of Chromatography A*, The Role of Theory in Chromatography, 1126 (1–2): 143–194. doi:10.1016/j.chroma.2006.06.074.

Wang, X. and P. W. Carr. 2007. An unexpected observation concerning the effect of anionic additives on the retention behavior of basic drugs and peptides in reversed-phase liquid chromatography. *Journal of Chromatography A* 1154 (1–2): 165–173. doi:10.1016/j.chroma.2007.03.057.

Weber, S. G. 1989. Theoretical and experimental studies of electrostatic effects in reversed-phase liquid chromatography. *Talanta* 36 (1): 99–106. doi:10.1016/0039-9140(89)80084-0.

Weber, S. G. and J. D. Orr. 1985. Establishment and determination of interfacial potentials and stationary phase dielectric constant in reversed-phase liquid chromatography. *Journal of Chromatography A* 322: 433–441. doi:10.1016/S0021-9673(01)97709-X.

Wittmer, D. P., N. O. Nuessle, and W. G. Haney. 1975. Simultaneous analysis of tartrazine and its intermediates by reversed phase liquid chromatography. *Analytical Chemistry* 47 (8): 1422–1423. doi:10.1021/ac60358a072.

2 Quantitative Mass Spectrometry Imaging of Molecules in Biological Systems

Ingela Lanekoff and Julia Laskin

CONTENTS

2.1 INTRODUCTION TO MASS SPECTROMETRY IMAGING

Mass spectrometry imaging (MSI) has been extensively used for obtaining spatial distributions of endogenous molecules (e.g., lipids, metabolites, and proteins), as well as drugs and their metabolites in biological samples. When combined with soft ionization techniques, MSI generates two- and three-dimensional maps of intact molecules extracted from specific locations on the sample. In a typical MSI experiment, a mass spectrum containing hundreds of features corresponding to analyte molecules extracted from the sample is obtained in each pixel. Ion images are generated by plotting ion intensities in individual mass spectra as a function of location on the sample. Because signal intensities in mass spectra are affected by several factors, including ionization efficiency of the analyte molecules, analyte concentration, and sample composition, quantification in MSI experiments is challenging. In particular, matrix effects defined as signal suppression during direct ionization of complex mixtures may affect the observed ion images and complicate quantification. Nevertheless, several strategies have been developed for both relative and absolute quantification in MSI. This review summarizes the current status of quantitative MSI (Q-MSI) and discusses future directions in this exciting field. The discussion will focus on several soft ionization techniques, where substantial effort has been dedicated to developing Q-MSI approaches.

By far, matrix-assisted laser desorption ionization (MALDI) is the most broadly used soft ionization technique in MSI experiments. MALDI imaging sources are commercially available and have been coupled to several mass analyzers. Time-of-flight (TOF) analyzers are particularly attractive because of their relatively low cost, good mass accuracy, fast data acquisition, and almost unlimited mass range. Both low- ($m/\Delta m$ < 5,000) and high-resolution ($m/\Delta m$ ~ 40,000) TOF instruments have been used in MALDI-MSI experiments. Meanwhile, high-resolution Orbitrap and Fourier transform ion cyclotron resonance (FT-ICR) instruments provide the advantages of excellent mass separation ($m/\Delta m$ > 60,000) at the expense of acquisition rate. High-resolution mass analysis is important for separating isobaric species commonly observed in MSI of chemically complex biological samples. Meanwhile, high mass accuracy facilitates peaks assignments in mass spectra. MALDI-MSI experiments have been performed both in vacuum and under ambient conditions with a typical spatial resolution of ~100 μm. Although somewhat less sensitive than vacuum MALDI, atmospheric pressure (AP)-MALDI eliminates the need of sample introduction into a vacuum system and is more readily adapted to different mass spectrometers [1]. AP-MALDI with a spatial resolution of better than 5 μm has been used to generate histology-like molecular images of tissue sections [2]. Similarly, AP-MALDI combined with electrospray ionization (MALDESI) has been used for imaging biological samples under ambient conditions [3].

Despite the wide use of MALDI-MSI, the requirement of matrix application prior to analysis, which is undesirable for some applications, has prompted development of ambient ionization techniques [4–7]. Ambient ionization enables analysis of biological samples without special sample pretreatment, which simplifies sample preparation protocols and eliminates the possible effect of matrix application on the localization of molecules in the sample [4–6,8–11]. A broad range of ambient ionization techniques developed in the past decade can be classified into methods that rely either on liquid-extraction or involve more energetic desorption of analytes from the sample using laser or plasma bombardment [12]. In this review, we will limit our discussion to liquid-extraction-based ambient ionization techniques [13].

Desorption electrospray ionization (DESI) is one of the most popular liquid-extraction ambient ionization techniques that has been successfully coupled with MSI. In DESI, the sample is bombarded with charged droplets accelerated toward the sample surface via a sheath gas flow. This process generates secondary charged droplets that contain analyte molecules extracted from a specific location on the sample. Subsequent desolvation of the secondary droplets generates ions corresponding to intact molecules present in the sample. Several variants of DESI have been developed to improve this technique's sensitivity to specific classes of analytes and its spatial resolution. The best spatial resolution of 35 μm has been reported in DESI-MSI of mouse brain tissue sections [14].

Another family of ambient ionization techniques relies on direct extraction of molecules into a liquid bridge formed between a specially designed sampling probe and the sample surface [13]. Different configurations of direct liquid-extraction probes have been summarized in a recent review and, hence, will not be discussed here [13]. Instead, we will focus on two direct liquid-extraction ambient ionization techniques that have been used in Q-MSI experiments. Specifically, we will limit our discussion to liquid-extraction surface analysis (LESA) [15] and nanospray desorption electrospray ionization (nano-DESI) [16] mass spectrometry. In LESA, analyte is first extracted into a static small droplet placed on the sample surface and is subsequently placed in front of a mass spectrometer for nanoelectrospray ionization. In contrast, in nano-DESI, the analyte is extracted into a flowing solvent delivered to the sample by a primary fused silica capillary and is removed through a secondary capillary that transfers the extracted analyte to a mass spectrometer inlet. The two capillaries comprise the nano-DESI probe. The difference in the respective probe designs translates directly into the differences in spatial resolution. Specifically, the typical spatial resolution of LESA is ~1 mm [17], whereas the best spatial resolution of ~350 μm has been achieved using an electrofocusing LESA probe [18]. In contrast, the typical spatial resolution of nano-DESI MSI is ~100–150 μm [19–22], whereas the best reported value is ~12 μm [23].

Quantification strategies discussed in this review rely on using standards of known concentration. In Q-MSI experiments, one or more standards are either deposited directly onto the sample or delivered to the sample via the extraction solvent. In Sections 2.3 and 2.4, we will provide a detailed summary of these approaches and will highlight some of the key applications of Q-MSI in biological research. Q-MSI is a rapidly growing field poised to transform MSI into an analytical technique that will enable accurate measurement of chemical gradients of hundreds of molecules in biological samples.

2.2 MATRIX EFFECTS IN MASS SPECTROMETRY IMAGING

Q-MSI experiments rely on the assumption that signal intensities in mass spectra obtained in individual pixels are proportional to analyte concentrations at each location. We begin by discussing *matrix effects* that may challenge this assumption and may affect ion images obtained in MSI experiments. Matrix effects, also referred to as ion suppression effects, occur during ionization of analyte mixtures as a result of molecules competing for charge [24–29]. Matrix effects occur in all ionization techniques and, as such, are not unique to MSI. They also affect ionization in liquid chromatography–mass spectrometry (LC–MS) experiments [30]. However, due to the nature of MSI, where all molecules from a specific location on the sample surface are ionized simultaneously, matrix effects originating from the sampled microenvironment may be more severe than in LC–MS. Matrix effects typically are studied and compensated for using an internal standard. Different approaches for applying internal standards will be discussed in more detail in the sections herein describing quantification. In Sections 2.2.1 and 2.2.2, we will discuss two types of matrix effects that are particularly important in MSI: (1) ionization suppression due to the analyte mixture's molecular composition and (2) variation in ionization efficiency resulting from differences in the alkali metal concentrations in the sample.

2.2.1 MATRIX EFFECTS DUE TO THE MOLECULAR COMPOSITION

During competitive ionization, molecules having a high affinity for charge suppress ionization of molecules with low affinity for charge. Therefore, the analyte mixture's molecular composition may significantly influence signal intensities obtained for individual compounds. As a result, signal intensities may not accurately reflect the concentration on the analyte molecules in the sample. The presence of matrix effects in MSI has been reported for different ionization techniques and various sample types [26–28,31–52]. The primary conclusion of these reports is that it is possible to compensate for matrix effects during MSI by incorporating a carefully selected internal standard. To compensate for matrix effects, the internal standard must have the same ionization efficiency as the target analyte. For example, Prideaux et al. used the internal standard levofloxacin homogenously sprayed over the tissue section to accurately determine the distribution of the drug moxifloxacin in rabbit lung tissue using MALDI-MSI [44]. The resulting ion image of the internal standard levofloxacin, shown in Figure 2.1b, displays distinct regions of higher intensity, indicating differences in matrix effects across the lung tissue section. Figure 2.1c shows the ion image of the administered drug, moxifloxacin, affected by matrix effects. By normalizing the intensity of moxifloxacin to the intensity of the internal standard levofloxacin, an ion image free of matrix effects is generated. This ion image, shown in Figure 2.1d, provides an accurate representation of the distribution of moxifloxacin in rabbit lung [44]. In another study, the distribution of levofloxacin in lung tissue post administration was found to be similar to the distribution of moxifloxacin. This result was obtained using deuterated levofloxacin as an internal standard added to the MALDI matrix solution [47]. In a related study, Pirman et al. used an internal standard to study organ-specific matrix effects in MALDI-MSI of acetylcarnitine [53] and found that different chemical compositions extracted from various organs had a pronounced impact on the internal standard's observed intensity [53]. Similarly, Rosen et al. examined tissue-specific matrix effects by imaging a thin tissue section of neonatal rat using infrared (IR)-MALDESI [39].

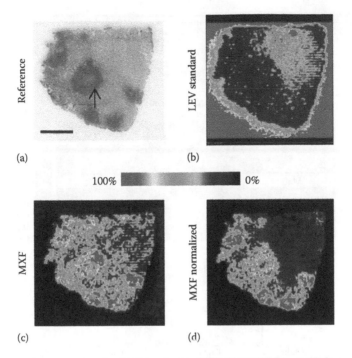

FIGURE 2.1   Figure showing ion images obtained by MALDI-MSI at each stage of the normalization process: (a) H&E reference image, (b) ion image of levofloxacin (LEV) standard, and (c) ion image of administered moxifloxacin (MXF). In (d), MXF normalized to LEV signals are observed from the granulomas (large central granuloma, indicated by the arrow in the H&E image) compared to that of the surrounding normal lung tissue. Greater LEV signal suppression occurred in the viable granuloma compared to that in the caseum, identifiable as the light pink center of the large central granuloma. Ion signal intensities were individually scaled for each image. Scale bar = 5 mm. (Reprinted with permission from Prideaux, B. et al., *Anal. Chem.*, 83, 2112–2118. Copyright 2011 American Chemical Society.)

Dong et al. investigated different strategies for introducing internal standards into DESI-MSI experiments [32]. They used glutaric acid as an internal standard to visualize the distribution of structurally similar organic acids in grapevine stem. Figure 2.2a shows ion images obtained for a stem that was soaked in the internal standard prior to DESI analysis, whereas ion images shown in Figure 2.2b were obtained by adding the internal standard to the DESI solvent during analysis. The top and middle rows of Figure 2.2a and b show ion images of the internal standard and the analytes normalized to the total ion current (TIC) that include matrix effects. The third row shows ion images of the analytes normalized to the internal standard, which compensates for matrix effects. The normalized ion images shown in Figure 2.2a and b differ strikingly, indicating that the approach used for incorporating the internal standard may affect the observed ion distributions. The authors suggest that if validation is not performed, differences in specific physicochemical interactions over the tissue surface may skew the results and cause misinterpretation of the obtained data [32].

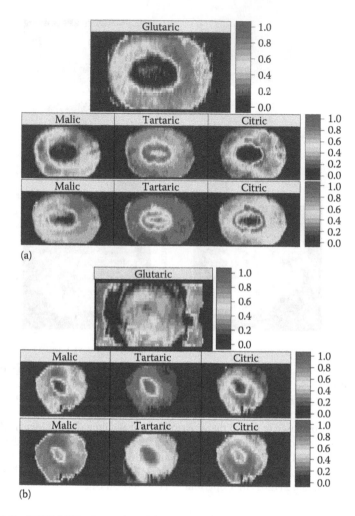

FIGURE 2.2 DESI-MSI of organic acids in grapevine stem using glutaric acid as the internal standard. Top rows show the ion image of the internal standard, middle rows feature ion images of three organic acids, and the bottom rows depict the ion images of the organic acids normalized to the internal standard. In (a), the stem was soaked in the internal standard prior to MSI, and in (b), the internal standard was included in the DESI solvent during MSI. (From Dong, Y. et al.: Impact of tissue surface properties on the desorption electrospray ionization imaging of organic acids in grapevine stem. *Rapid Commun. Mass Spectrom.* 2016. 30. 711–718. Copyright Wiley-VCH Verlag GmbH & Co. KGaA. Reproduced with permission.)

Matrix effects also have been investigated using nano-DESI [33,36,42,54]. In these studies, an internal standard is added to the nano-DESI solvent and supplied at a constant rate throughout the imaging experiment. Nonuniform distributions of nicotine-d3 used as a standard were observed in nano-DESI MSI of rat brain tissue sections, indicating the presence of matrix effects. Before normalization to

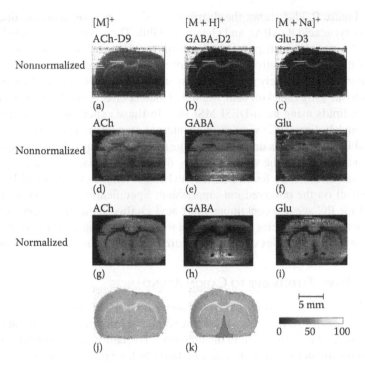

FIGURE 2.3 Ion images of ACh, GABA, and Glu in a rat brain tissue section. (a) Nonnormalized ion image of $[ACh-D_9]^+$, (b) nonnormalized ion image of $[GABA-D_2 + H]^+$, (c) nonnormalized ion image of $[Glu-D_3 + Na]^+$, (d) nonnormalized ion image of endogenous $[ACh]^+$, (e) nonnormalized ion image of endogenous $[GABA + H]^+$, (f) nonnormalized ion image of endogenous $[Glu + Na]^+$, (g) normalized ion image of $[ACh]^+$, (h) normalized ion image of $[GABA + H]^+$, (i) normalized ion image of $[Glu + Na]^+$, (j) optical image of the analyzed brain tissue section, and (k) optical image of analyzed brain tissue section with white matter regions highlighted in yellow and the medial septum-diagonal band complex highlighted in orange. Scale bar: 5 mm. The signal intensity of the ion images scale from dark to bright. (From Bergman, H.-M. et al., *Analyst*, 141, 3686–3695, 2016. Reproduced by permission of The Royal Society of Chemistry.)

the internal standard, nicotine signal was enhanced in the white matter regions, whereas normalization to the nicotine-d3 signal revealed a significantly different image free of matrix effects. This image revealed enhanced abundance of nicotine in the cortex and dentate gyrus [33]. Bergman et al. used nano-DESI MSI to investigate the impact of matrix effects on the distribution of small endogenous neurotransmitters in thin rat brain tissue sections by adding deuterated internal standards to the nano-DESI solvent [36]. Figure 2.3a–c shows ion images of the deuterated standards, acetylcholine (ACh-D9), γ-aminobutyric acid (GABA-D2), and glutamate (Glu-D3). Despite the uniform supply of the internal standards, the ion images show regions of increased intensities, suggesting reduced matrix effects in these regions. Figure 2.3d–f shows the distribution of endogenous acetylcholine (ACh), γ-aminobutyric acid (GABA), and glutamate (Glu). These ion images include matrix effects. The ion images displayed in the third row, Figure 2.3g–i, are normalized to the intensity of the respective internal standard in every pixel. Free of matrix effects, these ion images accurately represent the spatial localization of neurotransmitters in rat brain tissue [36]. Lanekoff et al. examined matrix effects involved during imaging of phospholipids using nano-DESI MSI [43]. In these experiments, two nonnatural phosphatidylcholines (PC) were used as internal standards. The native distribution of several PC species was determined by normalizing ion signals observed for these endogenous species to the signal intensity of the appropriate adduct of one of the internal standards. They found that the acyl chain length of individual PC species had an effect on the observed ion suppression. Specifically, lower ion signals were obtained for PC species containing short acyl chains, indicating more substantial suppression of these species in comparison to PC containing longer acyl chains during ionization of a complex chemical mixture extracted from the tissue sample [54].

2.2.2 MATRIX EFFECTS DUE TO CATION ABUNDANCES

Although matrix effects resulting from variations in alkali metal concentrations in biological samples have received less attention, the presence of such ion suppression effects has been discussed in the literature. By nature, the distribution of alkali metal cations in biological tissues may be fairly heterogeneous. Furthermore, tissue cryosectioning prior to MSI may introduce additional heterogeneity that originates from the plane through which the cells are sectioned [55]. Cross-sectioned cells typically display high amounts of potassium, whereas intact cells exhibit higher levels of sodium contained in the outer plasma membrane [55]. Because of the relatively high concentration of sodium and potassium in tissue sections, $[M + Na]^+$ and $[M + K]^+$ ions often dominate mass spectra obtained in MSI experiments. Of note, the preference for protonation, cationization on sodium, and cationization on potassium varies between molecular classes, which determines the relative abundance of different adducts produced during ionization. Naturally occurring variations in cation abundances on the sample surface may influence the observed signal intensities of different adducts in MSI. Recent advancements in MSI instrumentation enable imaging experiments with high spatial resolution approaching subcellular level (greater than 10 µm) [2]. Under these conditions, the sample surface's local microenvironment may increase pixel-to-pixel variations in the ion image because of the differences in matrix effects. For example, a spectrum acquired from a larger area ($\sim$100 × 100 µm^2) is an average of approximately 25 cells, and the plane through which individual cells have been sectioned may not have any substantial impact on the observed signal. However, imaging with higher spatial resolution of about 10 µm enables sampling of individual cells in each pixel. This may result in substantial scan-to-scan variations due to the increased heterogeneity of the chemical composition examined in each scan. Therefore, it is particularly important to account for matrix effects in MSI experiments performed with high spatial resolution.

To minimize the influence of matrix effects resulting from variations in cation abundances in tissue sections, Cerruti et al. added lithium to the uniformly applied matrix solution prior to MALDI-MSI [56]. When lithium concentration greatly exceeds the concentration of other cations in the sample, all molecules are preferentially cationized on lithium. They demonstrated that the exclusive cationization on lithium reduces matrix effects resulting from local variations of cation concentrations in the sample [56]. A second reported strategy to compensate for varying cation concentrations over the sample surface involves desalting tissue sections prior to MALDI-MSI [57–59]. Desalting is performed by drip-washing the sample in ammonium acetate followed by vacuum drying and matrix application [57]. To eliminate the effect of increased sodium and decreased potassium concentration in the ischemic region on results obtained in MALDI-MSI experiments, Wang et al. used desalted brain tissue sections of ischemic stroke models [57]. Lanekoff et al. developed a strategy that allows compensating for matrix effects in nano-DESI MSI resulting from heterogeneity in cation abundances by adding internal standards to the nano-DESI solvent and normalizing ion signals obtained for endogenous phospholipids to those of the corresponding standard's alkali metal adducts [42]. Figure 2.4 shows four endogenous PC species and their distributions in brain tissue sections of a mouse model of ischemic stroke. Figure 2.4a and b depicts ion images of sodium, $[M + Na]^+$, and potassium adducts, $[M + K]^+$, of these PC species, respectively. These ion images show an enhanced abundance of $[M + Na]^+$ and

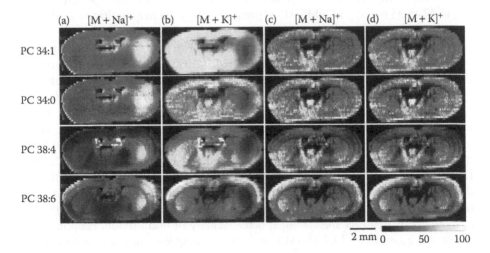

FIGURE 2.4 Ion images of sodium and potassium adducts, $[M + Na]^+$ and $[M + K]^+$, of endogenous PC 34:1, PC 34:0, PC 38:4, and PC 38:6. Ion images are normalized to the total ion current (TIC) in columns (a) and (b) and to the internal standard PC 25:0 in columns (c) and (d). Lateral scale bar is 2 mm. Intensity scale bar ranges from 0 (black) to 100% (light yellow) signal intensity of an individual peak. (From Lanekoff, I. et al., *Analyst*, 139, 3528–3532, 2014. Reproduced by permission of The Royal Society of Chemistry.)

reduced abundance of $[M + K]^+$ ions in the ischemic region, which is attributed to matrix effects. Figure 2.4c and d shows the same ion images after normalization to the internal standard, PC 25:0 (total number of carbons:total number of double bonds), using the appropriate alkali cation adduct. Similar ion images obtained for $[M + Na]^+$ and $[M + K]^+$ species after normalization demonstrate efficient compensation for matrix effects, which helps reveal the true distribution of various PC species in the sample [42].

2.3   QUANTITATIVE-MASS SPECTROMETRY IMAGING

Compensation for matrix effects is an important prerequisite for Q-MSI experiments that enable simultaneous analysis of the amount and localization of compounds in biological samples, including thin tissue sections. In addition, an accurate determination of analyte desorption efficiency from different locations on the sample is required for converting the observed signal intensity into both relative and absolute amounts of the analyte present in the sample. Both matrix effects and desorption efficiencies are influenced by the physical properties and chemical composition of the sample and must be taken into account in every pixel for each analyte ion. An additional challenge, specific to MALDI-MSI, is to ensure high quality and uniformity of the applied MALDI matrix resulting in formation of small crystals on the sample surface. For the interested reader, specific challenges for MALDI-MSI have been extensively discussed in several recent reviews [26,27,29,38,46,50,60–63].

LC–MS or LC–MS/MS experiments are often used to validate Q-MSI results [31,44,47,48,55,64–75]. LC–MS/MS is preferentially used in quantitative studies to ensure that isobaric species do not interfere with quantification. For the same reason, MSI uses MS/MS and/or mass spectrometers with high mass resolving power for quantitative analyses. When combining MS/MS-based quantification and an isotopically labeled standard, a wide isolation window, which enables simultaneous detection of fragment ions of both the analyte and standard in the same scan, typically is used [31,38,65]. Reyzer et al. quantified drug candidates in brain tissue using LC–MS/MS and compared the results with the intensities acquired using MALDI-MSI [65]. They concluded that the intensities obtained by MALDI-MSI reflected the relative amount of the drug in tissue as determined by LC–MS/MS. However, they proposed that because of the chemical heterogeneity of brain tissue, the intensities obtained from a single thin tissue section may not correlate with results obtained from a homogenate of the whole brain analyzed by LC–MS/MS [65]. To compensate for the chemical heterogeneity of brain tissue, Hankin et al. performed quantitative LC–MS/MS analysis of PC species in selected regions of rat brain tissue extracted using microdissection [55]. The intensities of four abundant PC species acquired using MALDI-MSI then were compared with the results of LC–MS/MS experiments. After the data were normalized to the intensity of the PC 34:1, a reasonable agreement between MALDI-MSI and LC–MS/MS was obtained [55]. Chumbley et al. investigated how different applications of an internal standard impacted the quantitative correlation of MALDI-MSI to LC–MS/MS. The best correlation of Q-MSI and quantitative LC–MS/MS was obtained when matrix application was performed after applying standards onto the tissue section [31]. Several other techniques have also been used for developing and validating Q-MSI. For example, gas chromatography (GC) [50], Raman spectroscopy imaging [76], surface plasmon resonance imaging [77], magnetic resonance imaging [78], positron emission tomography [79], and autoradiography [80–82] have been correlated with MSI. The term *Q-MSI* is interchangeably used to describe experiments involving either relative or absolute quantification (described in more detail in the following sections).

2.3.1 RELATIVE QUANTIFICATION WITH MASS SPECTROMETRY IMAGING

Relative Q-MSI typically refers to an experiment, where the absolute ion intensity of a compound of interest in each pixel of the ion image is converted into the relative abundance of the compound. This approach facilitates comparison between different samples or regions in the sample [83]. Relative Q-MSI is achieved either by defining compound-specific response factors (discussed in more detail later in this review) [45,51,72,84] or by introducing internal standards [32,35,36,69,72,75,85,86] to compensate for matrix effects [60]. Both strategies have been applied successfully for relative quantification. Using MALDI-MSI, Stoeckli et al. performed relative quantification by determining response factors for a drug candidate in different organs of a whole body tissue section [45]. Subsequently, organ-specific response factors were used for relative quantification of the drug candidate in different organs after *in vivo* administration (Figure 2.5) [45]. A second strategy for

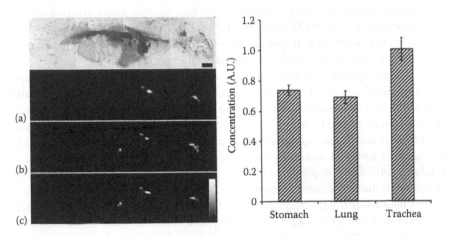

FIGURE 2.5 Relative quantification by MSI. Left: optical image (top) and MSI images from three different sections of the same animal (a–c). Data evaluation was obtained by averaging pixel intensities over the respective tissues. Error bars for the three samples appear at right. (Reprinted from *Int. J. Mass Spectrom.*, 260, Stoeckli, M., Staab, D., Schweitzer, A., Compound and metabolite distribution measured by MALDI mass spectrometric imaging in whole-body tissue sections, 195–202. Copyright 2007, with permission from Elsevier.)

relative quantification is to normalize the analyte signal intensity to that of deuterated standards. For relative quantification of cocaine and cocaine metabolites in hair strands, Porta et al. used deuterated internal standards sprayed onto the sample prior to MALDI-MSI [72]. The relative concentration of cocaine and its metabolites at different distances from the scalp were determined using the ratio of the ion intensity observed for each compound to the intensity of the respective deuterated standard [72]. Similarly, Schultz et al. reported relative quantification of the drug dasatinib in kidney tissue sections of differently treated animals using MALDI-MSI and a homologous distributed deuterated internal standard [69]. The intensity ratio of the ion signals obtained for dasatinib to the deuterated analog in one region of the tissue section was divided by the intensity ratio of the drug's total signal in the whole tissue section to the total signal of the deuterated analog in the whole tissue section. This approach eliminated differences in matrix effects and desorption efficiencies between different tissue sections, thereby enabling relative quantification [69].

2.3.2 ABSOLUTE QUANTIFICATION WITH MASS SPECTROMETRY IMAGING

Experiments in which signal intensities in individual pixels of the ion image are converted into absolute abundances of specific analyte molecules reported in grams or moles per gram tissue are typically referred to as *absolute Q-MSI*. In absolute Q-MSI, quantification is performed using an external calibration curve [60,83]. The analysis generally is performed by spotting calibration standards at different concentrations onto control tissue sections followed by MSI to generate an external standard calibration curve. The curve subsequently is used to convert the acquired signal intensity into analyte concentration at a particular location [44,49,50,53,61,66,70,71,79,87,88]. The weight of a single tissue section is then used to extrapolate the obtained concentrations to amount per tissue weight. However, because of the natural chemical heterogeneity of tissue samples, this approach does not necessarily account for matrix effects during MSI [31]. Källbeck et al. investigated how external standard curves, obtained after spotting calibration standards on a lung tissue section, were affected by using different normalization strategies [49]. Figure 2.6 shows the calibration curves obtained by engaging different normalization approaches evaluated in that study. Improved quantification was obtained using a deuterated internal standard, uniformly deposited onto the tissue section. This improvement was attributed to more efficient compensation for matrix effects originating from the tissue's molecular composition and suggests this was the best strategy for accurate quantification [49]. Lagarrigue et al. performed absolute quantification of a pesticide in mouse liver using an isotopically labeled standard and correlated the results to GC-electron capture detector (ECD) [50]. Bokhart et al. performed absolute quantification of the drug emtricitabine in tissue using IR-MALDESI by normalizing the analyte signal to that of the internal standard [71]. The following sections will describe several successful strategies for both online and offline quantification using internal standards.

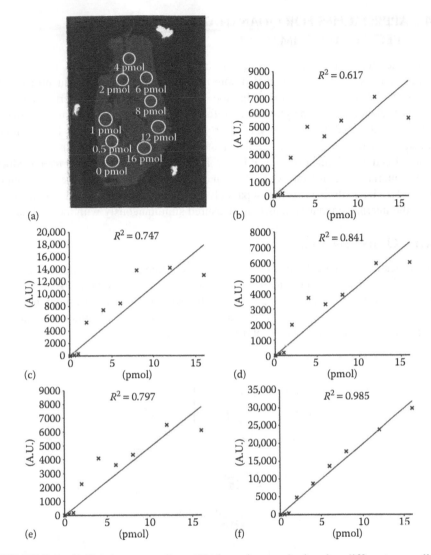

FIGURE 2.6 Calibration curve plots of imipramine standards using different normalization methods. (a) Digital photograph of a lung tissue section shown with different amounts of imipramine (blank, 0.5, 1, 2, 4, 6, 8, 12, and 16 pmol in 0.2 µl) applied on a rat lung tissue section. The average diameter of the spots was calculated to 1.51 mm. The correlation coefficients (R^2) for the concentration standards were calculated using different normalization methods: (b) nonnormalization, (c) median normalization, (d) TIC normalization, (e) root mean square (RMS) normalization, and (f) labeled normalization. Labeled normalization generated the best fit ($R^2 = 0.985$). The other normalization methods were better suited to nonlinear curves, such as logarithmic regression. The correlation coefficients obtained for logarithmic regression were nonnormalization, 0.902; median normalization, 0.926; TIC normalization, 0.868; RMS normalization, 0.937; and labeled normalization, 0.899. (Reprinted from *J. Proteomics*, 75, Kallback, P. et al., Novel mass spectrometry imaging software assisting labeled normalization and quantitation of drugs and neuropeptides directly in tissue sections, 4941–4951. Copyright 2012, with permission from Elsevier.)

2.4 APPROACHES FOR QUANTITATIVE-MASS SPECTROMETRY IMAGING

Approaches for Q-MSI involve using one or more internal standards, usually iso-topically labeled analogs or unlabeled homologs of the analyte to be quantified. An isotopically labeled internal standard has the same physiochemical properties and ion-ization efficiency as the analyte and, therefore, experiences the same matrix effects as the extracted analyte molecules. Similarly, homologous compounds have similar ionization efficiency as analytes of interest and are assumed to experience comparable suppression during ionization. Several approaches have been developed for introduc-ing the internal standard into Q-MSI, including offline strategies, where the inter-nal standard intensity is measured separately from the analyte, and online strategies, where the internal standard intensity is measured simultaneously with the analyte.

2.4.1 OFFLINE STRATEGIES

Offline strategies can be divided into the following three categories (shown schemat-ically in Figure 2.7): (1) applying the internal standard onto a separate tissue section to obtain a calibration curve, (2) applying the internal standard onto a separate tissue section to determine response factors, and (3) incorporating the internal standard into a homogenized tissue that is solidified and sectioned.

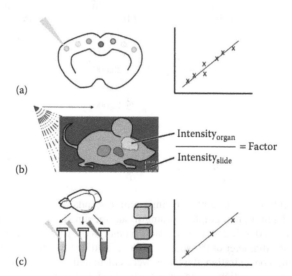

FIGURE 2.7 Schematic illustration of three strategies for offline Q-MSI using standards. (a) Standard is spotted onto a tissue section at several concentrations, and the acquired intensi-ties are used to generate an external calibration curve. (b) Standard is sprayed uniformly onto a tissue section and glass slide, and the acquired intensities in one organ are normalized to the intensity of the standard on the glass slide to determine a tissue-specific response factor. (c) Tissue is homogenized, and standard is added to the homogenate at several concentrations. The homogenate is solidified and sectioned, and the acquired intensities are used to generate an external calibration curve.

2.4.1.1   Spotting Standard onto Separate Tissue Section

In the approach illustrated in Figure 2.7a, spots containing different concentrations of a standard are deposited directly onto a tissue section. MSI is used to generate an external calibration curve, which correlates the standard concentration to the detected signal intensity in the complex chemical matrix. Several studies have generated external standard curves of drugs on control tissue using MALDI-MSI then quantified the amount of drug present in tissue after drug administration [49,66,79,87]. In these studies, the drug itself is often used as a standard as it is not present in the control tissue. For example, Hsieh et al. spotted different concentrations of a solution containing two standards, namely the drug clozapine and its metabolite norclozapine, on top of brain tissue to generate external calibration curves for both species using MALDI-MSI [87]. A similar approach was employed for quantifying the amount of tiotropium and imipramine in lung tissues after drug administration [49,66] and for evaluating the average concentration of two drugs in kidney tissue sections following dosing [79]. Despite the approach's success, it is difficult to generate a homogenous distribution of the standard in each spot due to a coffee-ring effect, which results in enhanced abundance of the standard at the edge of the spot [38,71,89]. To ensure high quality of the calibration spots in MALDI-MSI, Shroff et al. added fluorescent dye to the internal standard solution and determined the distribution of the standard in each spot using a fluorescence scanner [90]. Other studies have carefully selected the entire area of the spotted standard to get an average response, which helped to compensate for the coffee-ring effect [31].

2.4.1.2   Response Factors

Instead of generating calibration curves, tissue-specific response factors are often introduced to compensate for matrix effects in different parts of the tissue and to facilitate quantification in Q-MSI [45,74,84]. To find the organ where the administered drugs localize, the tissue-specific response approach has been applied to whole body sections. In these experiments, internal standards are incorporated into the sample by soaking the tissue section in solution containing the standard prior to MSI. Tissue-specific ionization efficiency factors have been derived from MSI of whole body sections by accounting for the observed signal suppression in different organs [45]. Similarly, Hamm et al. introduced tissue extinction coefficient (TEC) corresponding to signal intensity loss due to matrix effects in different organs of a whole body rat section [84]. After spraying standard solution on the sample, TEC is calculated using the intensity of the standard obtained on the tissue section divided by the intensity of the standard on the glass slide next to the tissue section. Dependence of TEC on the analyte concentration was examined by varying the standard's concentration. It was found that TEC does not depend on the standard's concentration, indicating that signal suppression is determined only by the composition of the complex mixture extracted from the tissue. To enable quantification, a calibration curve was generated by depositing standards onto the glass slide adjacent to the tissue. The drug concentration then was determined by dividing the average intensity of the drug obtained for each organ by the respective TEC [84]. Response factors have also been used for quantification in DESI-MSI. For example, Vismeh et al. deposited nine spots of loxapine calibration solution

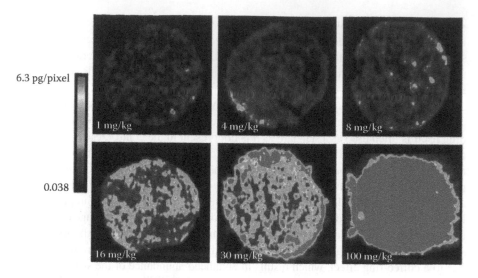

FIGURE 2.8 MSI images of olanzapine in liver obtained from animals dosed at 1, 4, 8, 16, 30, and 100 mg/kg. (From Koeniger, S. L. et al.: A quantitation method for mass spectrometry imaging. *Rapid Commun. Mass Spectrom.* 2011. 25. 503–510. Copyright Wiley-VCH Verlag GmbH & Co. KGaA. Reproduced with permission.)

onto clozapine-treated rat brain tissue sections [51]. To minimize pixel-to-pixel variations, response factors were calculated using the ratio of clozapine to the standard. Quantification was performed by generating an external calibration curve [51]. Koeniger et al. introduced a different approach and defined a conversion factor to help correlate quantification using MALDI-MSI and LC–MS/MS [74]. In this approach, adjacent liver tissue sections were analyzed using both LC–MS/MS and MALDI-MSI after drug administration at different concentrations. Drug concentrations obtained with LC–MS/MS were used to find a conversion factor between the obtained MALDI-MSI intensity, in total ion counts per tissue sections and drug concentration in the tissue. Figure 2.8 shows representative quantitative ion images of the drug olanzapine in liver for each administered concentration [74]. Similar to tissue-specific ionization efficiency factors and TEC, the conversion factor is useful for quantifying average amounts of analyte in a tissue volume, such as an organ.

2.4.1.3 Incorporating Internal Standard into a Homogenized Tissue

Several groups have introduced tissue models to limit the amount of animal experiments required to explore the effect of drug dose on its accommodation in tissues. Typically, these models are generated using solidified tissue homogenates containing standards at different concentrations [67,73]. Groseclose et al. added standards of the drug lapatinib into tissue homogenate that was subsequently frozen into a polymer support mold and cryosectioned [67]. Using MALDI-MSI, an external calibration curve was generated by examining tissue models containing different concentrations of the drug. The calibration curve was then applied to estimate lapatinib concentrations in a dosed animal tissue section [67]. Similarly, to generate an external

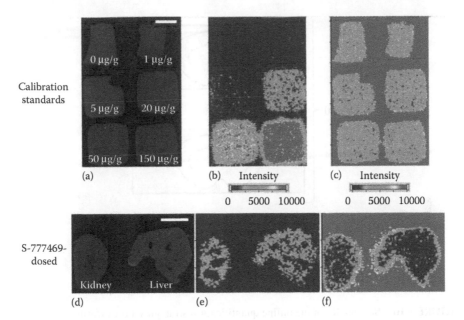

FIGURE 2.9   Representative optical images (a and d) and ion images of *m/z* 415 → 369 (b and e) and *m/z* 425 → 379 (c and f) on the sections of the calibration standards (a–c) or the kidney and liver (d–f). Bar = 55 mm. (Reproduced with permission from Takai, N. et al., *Mass Spectrom. (Tokyo)*, 3(1), A0025, 2014.)

calibration curve, Jadoul et al. added different amounts of deuterated PC to homogenized brain tissue [73]. This curve helped quantify endogenous PC species in a mouse brain tissue section using MALDI-MSI. Although the entire brain was homogenized in that study, the authors proposed that homogenates of particular regions of the brain could be used to better account for region-specific matrix effects when analyzing heterogeneous brain tissues [73]. Takai et al. used the combination of a tissue model and an internal standard deposition onto tissue sections for Q-MSI of a drug candidate in liver [68]. By depositing internal standard onto the model tissue, the study generated a calibration curve displaying the ratio of the analyte to the internal standard against the concentration of the standard. Afterward, MALDI-MSI was used to image thin tissue sections of multiple organs containing a uniformly deposited internal standard. Figure 2.9a–c shows sections of the tissue model containing different concentrations of the drug candidate (Figure 2.9b) and the corresponding deuterated internal standard (Figure 2.9c). Figure 2.9d–f depicts images of dosed kidney and liver containing analyte (Figure 2.9e) and internal standard (Figure 2.9f) [68].

2.4.2 Online Strategies

Online quantification strategies rely on simultaneous detection of the internal standard and the extracted analyte. Several reports have shown that pixel-to-pixel signal variations are minimized by normalizing the analyte signal to that of the internal standard [33,35,36,42,49–51,53,54,65,70,71,91–93]. Furthermore, normalization to

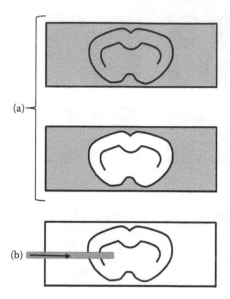

FIGURE 2.10 Schematics of the online quantification strategies. (a) Uniform deposition of an internal standard on top of or under the tissue section and (b) internal standard incorporated into the solvent used for liquid extraction.

the internal standard signal observed in every pixel helps account for matrix effects occurring in individual ionization events during MSI experiments. Successful online quantification relies on the following assumptions: (1) the internal standard is present at equal concentrations in every spectrum and (2) there is a close correspondence between the physiochemical properties of the internal standard and the analyte. The latter is achieved by using an isotopically labeled version of the analyte as an internal standard. Several strategies illustrated in Figure 2.10 have been employed to enable incorporation of internal standard(s) into Q-MSI experiments. These include (in Figure 2.10) (a) coating the internal standard either on top or under the tissue section and (b) incorporating internal standards into the extraction solvent.

2.4.2.1 Coating with Internal Standard

2.4.2.1.1 On Top of Tissue

A common online quantification strategy for MALDI-MSI involves depositing a uniform layer of an internal standard onto the tissue section in a similar manner as spraying the MALDI matrix. The internal standard can either be sprayed onto the tissue prior to spraying the MALDI matrix or incorporated in the MALDI matrix solution. Chumbley et al. evaluated different methods of applying standards for quantification in MALDI-MSI on tissues dosed with rifampicin (RIF) [31]. Specifically, they used a robotic spotter to reproducibly apply an array of ~200 μm diameter RIF microspots onto both control tissue sections and tissues dosed with RIF in vitro and in vivo. To ensure adequate averaging over individual microspots, imaging experiments were performed with a lateral resolution of 200 μm. The most accurate results, in comparison with LC–MS/MS, were obtained by depositing the standard on top of the tissue followed by matrix deposition.

MALDI-MSI of control tissues prepared using the same protocol enabled quality control validation of Q-MSI [31]. Similarly, the pesticide chlordecone was quantified in mouse liver post administration by including an isotopically labeled internal standard into the MALDI matrix solution [50]. In a related study, Reyzer et al. included an internal standard into the MALDI matrix [65]. Källback et al. quantified endogenous Substance P by spotting deuterated Substance P onto the cortex of the brain where the signal from endogenous Substance P was below the detection limit [49]. To quantify endogenous proteins in rat brain tissue sections, Clemis et al. sprayed a solution of MALDI matrix containing isotopically labeled peptides and trypsin onto the tissue section, which had been prewashed to remove lipids and metabolites [92]. Peptides produced by tryptic digestion of proteins were detected using MALDI-MSI. The intensity of the internal standard and endogenous peptide in each scan was used to construct a calibration curve and to compensate for matrix effects in individual pixels [92].

2.4.2.1.2   Under Tissue

Q-MSI experiments have also been performed using an internal standard applied uniformly under the tissue section. Then, the internal standard is extracted into the matrix crystals in a similar manner as the analyte within the tissue. Pirman et al. used deuterated acetylcarnitine as an internal standard and applied it under the tissue section to quantify endogenous acetylcarnitine in the sample [53,88]. A standard addition curve was generated by depositing acetylcarnitine at different concentrations onto the tissue section while accounting for matrix effects by normalizing to the deuterated internal standard [53]. Figure 2.11 shows the resulting ion images and standard curves for acetylcarnitine. Figure 2.11a displays the ion image of acetylcarnitine, illustrating both the endogenous and deposited standard at m/z 204. In Figure 2.11b, the ion images of deuterated acetylcarnitine display matrix effects originating from the chemical heterogeneity of the tissue section. Figure 2.11c shows the ion image obtained by normalizing the signal of acetylcarnitine to deuterated acetylcarnitine. Normalization to the deuterated standards minimized standard deviations in the calibration curves (shown in Figure 2.11d and e) [53,70]. Bokhart et al. quantified the HIV drug emtricitabine in cervical tissue using MALDESI by generating an external calibration curve of deposited isotopically labeled emtricitabine on top of the tissue section and lamivudine as an internal standard under the tissue [71]. Landgraf et al. quantified endogenous PC in tissue in each pixel using MALDI-MSI by uniformly depositing an internal standard under the tissue [85]. The amount of the observed analyte in moles/pixel was calculated using Equation 2.1:

$$\text{Moles} = \frac{\text{Int}_{AN}}{\text{Int}_{IS}} \times \frac{W_{IS}}{\text{MW}_{AN}} \times A \qquad (2.1)$$

where:
  Int_{AN} is the analyte signal intensity
  Int_{IS} is the intensity of the internal standard
  W_{IS} is the amount of the internal standard deposited onto the glass slide
  MW_{AN} is the analyte's molecular weight
  A is the area of the tissue (Figure 2.12) [85]

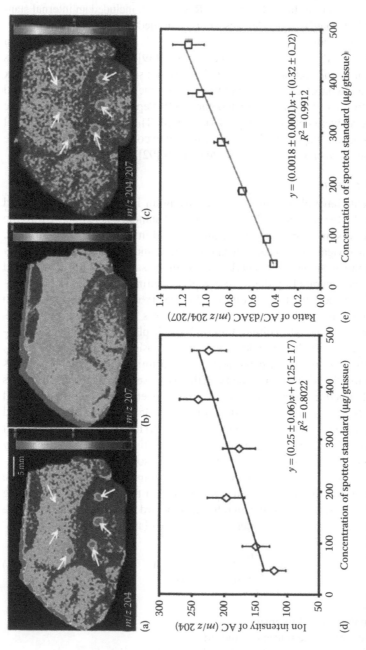

FIGURE 2.11 MS image of piglet brain tissue for the detection of endogenous acetylcarnitine. (a) Image of the [M + H]⁺ ion correspond ng to acetylcarnitine (m/z 204) shows a clear differentiation between the brain's white and gray matter. (b) Applied acetylcarnitine-d3 shows a similar pattern, mplying that there is either decreased extraction or ionization from the two tissue types. (c) Ratio image of acetylcarnitine/acetylcarnitine-d3 results in a better representation of the distribution of acetylcarnitine within the brain sample while improving the calibration curve's linearity. Calibration curves plotted with (d) the ion intensity of m/z 204 versus (e) the ratio of m/z 204/207. The appearance of saturated calibration spots is graphical due to scaling to show the endogenous levels of acetylcarnitine. (Reprinted with permission from Pirman, D. A. et al., *Anal. Chem.*, 85, 1090–1096. Copyright 2013 American Chemical Society.)

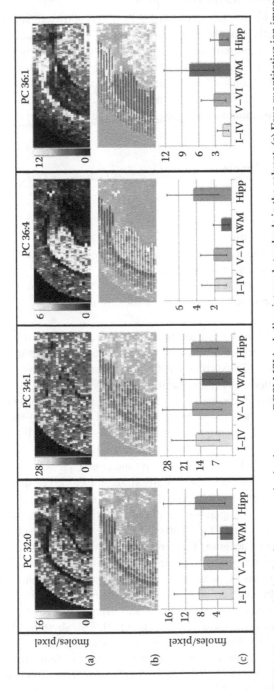

FIGURE 2.12 Quantitative ion images acquired using nano-DESI MSI including internal standards in the solvent. (a) Four quantitative ion images of endogenous PC. From left to right: PC 32:0, PC 34:1, PC 36:4, and PC 36:1 are displayed as [M + K]⁺ signals normalized to [PC 43:6 + K]⁺. The values in the intensity scale bar show fmol/pixel and vary between ion images. Scale bar shows 2 mm. (b) Regions of interest are color coded as: yellow, neocortical layers I–IV; green, neocortical layers V–VI; red, white matter (WM); and blue, hippocampus (Hipp). (c) Graphs showing the average amount of the corresponding PC in each region of the brain in fmoles per pixel (color coded as described in b). Error bars show standard deviations of individual pixels in the respective region. Number of pixels within each region: I–IV, $n = 58$; V–VI, $n = 356$; WM, $n = 249$; and Hipp, $n = 477$. (Reprinted with permission from Lanekoff, I. et al., *Anal. Chem.*, 86, 1872–1880. Copyright 2014 American Chemical Society.)

2.4.2.2 Quantitative-Mass Spectrometry Imaging Using Ambient Liquid-Extraction Techniques

Q-MSI studies using liquid-extraction techniques have the ability to incorporate the internal standard directly into the extraction solvent. This approach ensures that the internal standard is present at a constant concentration throughout an imaging experiment. By including a known concentration of deuterated nicotine into the nano-DESI solvent, Lanekoff et al. quantified nicotine in rat brain after administration [33]. This approach used the intensity ratio of nicotine to deuterated nicotine to calculate the amount nicotine detected in each pixel, according to Equation 2.2:

$$\text{Amount nicotine}\left(\frac{\text{moles}}{\text{pixel}}\right) = C_{std} \times \frac{I_{nicotine}}{I_{std}} \times \text{Flowrate}_{std} \times \text{IT} \qquad (2.2)$$

where:

$I_{nicotine}$ is the intensity of nicotine
I_{std} is the intensity of the deuterated nicotine used as an internal standard
C_{std} is the concentration of deuterated nicotine
Flowrate_{std} is the solvent flow rate
IT is the ion accumulation time for each scan in the Orbitrap [33]

In a related study, nano-DESI MSI was used to quantify three small neurotransmitters in a rat brain tissue section simultaneously by incorporating three deuterated internal standards into the nano-DESI solvent [36]. Internal standards have also been incorporated into the extraction solvent for quantitative analysis using LESA [52,94]. Parson et al. added deuterated chloroquine into LESA solvent and spatially profiled the amount of chloroquine in the kidneys of dosed rats [52]. Figure 2.13 shows that the highest concentration of chloroquine was found in the renal medulla. By also including deuterated internal standards of chloroquine metabolites in the solvent, the study followed chloroquine metabolism over a 24 h time course [52,95].

Ambient liquid-extraction techniques are ideally suited for Q-MSI of lipids in tissue sections. Specifically, Lanekoff et al. have developed a shotgun-like quantification approach for Q-MSI of all endogenous PC in rat brain tissue section using only two PC standards added to the nano-DESI solvent [43]. The approach used a modified version of Equation 2.2 that accounts for carbon factors representing differences in ion suppression stemming from differences in acyl chain length [43]. Carbon factors were determined using two PC standards: one with shorter and another with longer acyl chain lengths. In addition, the extraction efficiency of endogenous PC species from the sample surface was investigated and discovered to be consistent between different regions of the rat brain tissue section [43]. This shotgun-like quantification approach was used to obtain quantitative images for 22 PC species using only two standards. Figure 2.12a shows quantitative ion images of four different PC species, namely PC 32:0, PC 34:1, PC 36:4, and PC 36:1 with their individual intensity scales in fmoles/pixel. Figure 2.12c depicts the quantitative data represented by bar graphs from specific regions of interest defined in Figure 2.12b. This approach

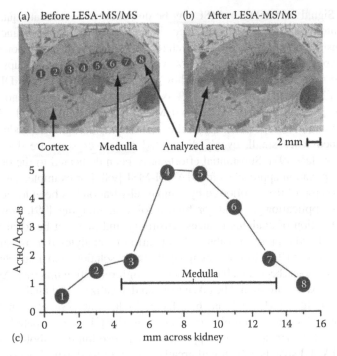

FIGURE 2.13 Spatial distribution of chloroquine (CHQ) in the kidney from a CHQ-dosed animal tissue section (10 mg/kg; 2 h). (a) Optical image of the kidney tissue section prior to LESA-MS/MS analysis. The numbered blue circles were created using LESA; clarity software (1 mm diameter with 2 mm center-to-center spacing) and selected LESA-MS/MS analysis points. (b) Optical image of kidney section after LESA-MS/MS analysis illustrating the actual areas analyzed across the cortex and medulla (outlined by red dashes in [a]). Each analysis spot was approximately 1.5–2 mm in diameter. (c) Integrated selected reaction monitoring signals of CHQ normalized to the internal standard, CHQ-d4, are plotted versus distance in mm across the kidney section. The numbered blue circles correlate with the numbered circles in (a) and indicate the intensity versus spatial location in the cortex (circles 1, 2, and 8) and the medulla (circles 3–7). (From Parson, W.B. et al.: Analysis of chloroquine and metabolites directly from whole-body animal tissue sections by liquid-extraction surface analysis (LESA) and tandem mass spectrometry. *J. Mass Spectrom.* 2012. 47. 1420–1428. Copyright Wiley-VCH Verlag GmbH & Co. KGaA. Reproduced with permission.)

may be readily expanded to enable simultaneous quantification of several classes of lipids by adding appropriate standards to the nano-DESI solvent.

2.5 FUTURE DIRECTIONS

Further development and validation of Q-MSI are needed to ensure its accuracy, robust operation, and applicability to a broad range of compounds. By adding several types of standards, it may be possible to obtain quantitative images of a large variety of endogenous species present in the sample. Several aspects affect Q-MSI accuracy, including sample preparation, pixel-to-pixel signal stability, and analyte extraction efficiency. Signal variability in MSI may be detrimental to accurate quantification. A recent interlaboratory study examined the repeatability and constancy of DESI signal intensities obtained for the same reference samples across 20 laboratories [96]. Although repeatability of better than 20% was obtained by several groups, imperfections in the source design have been shown to substantially degrade DESI performance. Similar repeatability of ~15–20% is commonly achieved in nano-DESI MSI experiments [23]. Shot-to-shot variations in MALDI experiments may also be reduced by improving matrix deposition approaches. For example, higher reproducibility has been obtained when small, evenly distributed matrix crystals are deposited onto the sample surface [97]. Substantial efforts have been dedicated to the optimization of matrix application approaches for MALDI-MSI [98]. For example, sublimation of the matrix on top of tissue followed by sample rehydration has been identified as the best matrix application method for high-resolution imaging [99]. However, less efficient extraction of analytes reduces sensitivity and may not be ideal for Q-MSI experiments. Furthermore, possible delocalization of analytes during sample rehydration is a concern [100]. Robotic spraying of matrix solution onto a sample is used by many groups and has been a viable approach for high-resolution imaging. Again, when using this approach, care should be taken to avoid delocalization of analytes. Moreover, signal repeatability in MALDI may be affected by laser ablation efficiency, which is controlled by adjusting the number of laser shots [101]. It is anticipated that further optimization of matrix application approaches will also improve shot-to-shot repeatability of MALDI signals, which will greatly benefit Q-MSI experiments.

Analyte extraction efficiency is another important factor affecting Q-MSI accuracy. As discussed, analyte extraction in MALDI is affected by the matrix application approach and the volume of ablated material controlled by the number of laser shots. It is known that tissue composition also may affect analyte extraction efficiency in MALDI [98]. Similarly, tissue composition could affect the extraction efficiency in DESI and direct liquid-extraction ambient ionization approaches. Although similar extraction efficiency has been observed from the white and gray matter in nano-DESI MSI of rat brain tissue sections [54], these observations cannot be extrapolated to other types of tissues. It follows that additional experiments are necessary to characterize the dependence of extraction efficiency on the tissue composition in Q-MSI experiments.

In summary, Q-MSI is a powerful technique for mapping chemical gradients in complex biological systems. Better understanding of the processes relevant to Q-MSI will assist in developing approaches for robust quantitative analysis of molecules in chemically complex biological samples. Compared to bulk analysis after tissue homogenization, MSI enables quantitative analysis of molecules present in relatively small regions of tissue sections, which may be difficult to detect in tissue homogenates. Q-MSI requires a targeted approach and relies on using carefully selected internal standards to account for tissue-specific matrix effects. The best strategy involves using isotopically labeled internal standards and normalizing the intensity of the analyte to the intensity of the internal standard in the same pixel to account for pixel-specific matrix effects. Shotgun-like quantification of lipids can also be performed using only two standards from the same lipid class. This approach can be readily expanded to other biomolecules and can be used for simultaneous Q-MSI of multiple analyte classes. Future advancements in the development and validation of robust Q-MSI approaches will establish this technique as a transformational tool for studies in biology, drug discovery, and clinical research.

ACKNOWLEDGMENTS

Ingela Lanekoff acknowledges support from the Swedish Research Council (621-2013-4231) and the Swedish Foundation for Strategic Research (ICA-6). Julia Laskin acknowledges support from the Chemical Imaging Initiative at Pacific Northwest National Laboratory (PNNL) and partial support from National Institutes of Health (NIH) grant R21ES024229-01. Early development of nano-DESI was supported by the U.S. Department of Energy (DOE), Office of Basic Energy Sciences, Division of Chemical Sciences, Geosciences & Biosciences. Research at PNNL (Julia Laskin) is performed at Environmental Molecular Science Laboratory (EMSL), a national scientific user facility sponsored by the DOE's Office of Biological and Environmental Research. PNNL is a multi-program national laboratory operated by Battelle for the DOE under Contract DE-AC05-76RL01830.

REFERENCES

1. Koestler, M. et al., A high-resolution scanning microprobe matrix-assisted laser desorption/ionization ion source for imaging analysis on an ion trap/Fourier transform ion cyclotron resonance mass spectrometer. *Rapid Communications in Mass Spectrometry*, 2008. **22**(20): 3275–3285.
2. Römpp, A. and B. Spengler, Mass spectrometry imaging with high resolution in mass and space. *Histochemistry and Cell Biology*, 2013. **139**(6): 759–783.
3. Sampson, J.S., A.M. Hawkridge, and D.C. Muddiman, Generation and detection of multiply-charged peptides and proteins by matrix-assisted laser desorption electrospray ionization (MALDESI) Fourier transform ion cyclotron resonance mass spectrometry. *Journal of the American Society for Mass Spectrometry*, 2006. **17**(12): 1712–1716.
4. Cooks, R.G. et al., Ambient mass spectrometry. *Science*, 2006. **311**(5767): 1566–1570.
5. Weston, D.J., Ambient ionization mass spectrometry: Current understanding of mechanistic theory; analytical performance and application areas. *Analyst*, 2010. **135**(4): 661–668.
6. Wu, C. et al., Mass spectrometry imaging under ambient conditions. *Mass Spectrometry Reviews*, 2013. **32**(3): 218–243.
7. Monge, M.E. et al., Mass spectrometry: Recent advances in direct open air surface sampling/ionization. *Chemical Reviews*, 2013. **113**(4): 2269–2308.
8. Badu-Tawiah, A.K. et al., Chemical aspects of the extractive methods of ambient ionization mass spectrometry. *Annual Review of Physical Chemistry*, 2013. **64**(1): 481–505.
9. Ifa, D.R. et al., Desorption electrospray ionization and other ambient ionization methods: Current progress and preview. *Analyst*, 2010. **135**(4): 669–681.
10. Van Berkel, G.J., S.P. Pasilis, and O. Ovchinnikova, Established and emerging atmospheric pressure surface sampling/ionization techniques for mass spectrometry. *Journal of Mass Spectrometry*, 2008. **43**(9): 1161–1180.
11. Hsu, C.-C. and P.C. Dorrestein, Visualizing life with ambient mass spectrometry. *Current Opinion in Biotechnology*, 2015. **31**: 24–34.
12. Venter, A.R. et al., Mechanisms of real-time, proximal sample processing during ambient ionization mass spectrometry. *Analytical Chemistry*, 2014. **86**(1): 233–249.

13. Laskin, J. and I. Lanekoff, Ambient mass spectrometry imaging using direct liquid extraction techniques. *Analytical Chemistry*, 2016. **88**(1): 52–73.

14. Campbell, D.I. et al., Improved spatial resolution in the imaging of biological tissue using desorption electrospray ionization. *Analytical and Bioanalytical Chemistry*, 2012. **404**(2): 389–398.

15. Kertesz, V. and G.J. Van Berkel, Fully automated liquid extraction-based surface sampling and ionization using a chip-based robotic nanoelectrospray platform. *Journal of Mass Spectrometry*, 2010. **45**(3): 252–260.

16. Roach, P.J., J. Laskin, and A. Laskin, Nanospray desorption electrospray ionization: An ambient method for liquid-extraction surface sampling in mass spectrometry. *Analyst*, 2010. **135**(9): 2233–2236.

17. Swales, J.G. et al., Mass spectrometry imaging of cassette-dosed drugs for higher throughput pharmacokinetic and biodistribution analysis. *Analytical Chemistry*, 2014. **86**(16): 8473.

18. Brenton, A.G. and A.R. Godfrey, Electro-focusing liquid extractive surface analysis (EF-LESA) coupled to mass spectrometry. *Analytical Chemistry*, 2014. **86**(7): 3323–3329.

19. Abraham, J.L., S. Chandra, and A. Agrawal, Quantification and micron-scale imaging of spatial distribution of trace beryllium in shrapnel fragments and metallurgic samples with correlative fluorescence detection method and secondary ion mass spectrometry (SIMS). *Journal of Microscopy*, 2014. **256**(2): 145–152.

20. Lanekoff, I. et al., Three-dimensional imaging of lipids and metabolites in tissues by nanospray desorption electrospray ionization mass spectrometry. *Analytical and Bioanalytical Chemistry*, 2015. **407**(8): 2063–2071.

21. Lanekoff, I. et al., Automated platform for high-resolution tissue imaging using nanospray desorption electrospray ionization mass spectrometry. *Analytical Chemistry*, 2012. **84**(19): 8351–8356.

22. Lanekoff, I. and J. Laskin, Imaging of lipids and metabolites using nanospray desorption electrospray ionization mass spectrometry. *Methods in Molecular Biology (Clifton, N.J.)*, 2015. **1203**: 99–106.

23. Laskin, J. et al., Tissue imaging using nanospray desorption electrospray ionization mass spectrometry. *Analytical Chemistry*, 2012. **84**(1): 141–148.

24. Annesley, T.M., Ion suppression in mass spectrometry. *Clinical Chemistry*, 2003. **49**(7): 1041–1044.

25. Knochenmuss, R. et al., The matrix suppression effect and ionization mechanisms in matrix-assisted laser desorption/ionization. *Rapid Communications in Mass Spectrometry*, 1996. **10**(8): 871–877.

26. Ellis, S.R., A.L. Bruinen, and R.M.A. Heeren, A critical evaluation of the current state-of-the-art in quantitative imaging mass spectrometry. *Analytical and Bioanalytical Chemistry*, 2014. **406**(5): 1275–1289.

27. Lietz, C.B., E. Gemperline, and L. Li, Qualitative and quantitative mass spectrometry imaging of drugs and metabolites. *Advanced Drug Delivery Reviews*, 2013. **65**(8): 1074–1085.

28. Prideaux, B. and M. Stoeckli, Mass spectrometry imaging for drug distribution studies. *Journal of Proteomics*, 2012. **75**: 4999.

29. Castellino, S., M.R. Groseclose, and D. Wagner, MALDI imaging mass spectrometry: Bridging biology and chemistry in drug development. *Bioanalysis*, 2011. **3**(21): 2427–2441.

30. Taylor, P.J., Matrix effects: The Achilles heel of quantitative high-performance liquid chromatography-electrospray-tandem mass spectrometry. *Clinical Biochemistry*, 2005. **38**(4): 328–334.

31. Chumbley, C.W. et al., Absolute quantitative MALDI imaging mass spectrometry: A case of rifampicin in liver tissues. *Analytical Chemistry*, 2016. **88**(4): 2392–2398.
32. Dong, Y., G. Guella, and P. Franceschi, Impact of tissue surface properties on the desorption electrospray ionization imaging of organic acids in grapevine stem. *Rapid Communications in Mass Spectrometry*, 2016. **30**(6): 711–718.
33. Lanekoff, I. et al., Imaging nicotine in rat brain tissue by use of nanospray desorption electrospray ionization mass spectrometry. *Analytical Chemistry*, 2012. **85**(2): 882–889.
34. Muramoto, S. et al., Test sample for the spatially resolved quantification of illicit drugs on fingerprints using imaging mass spectrometry. *Analytical Chemistry*, 2015. **87**(10): 5444–5450.
35. Park, K.M. et al., Relative quantification in imaging of a peptide on a mouse brain tissue by matrix-assisted laser desorption ionization. *Analytical Chemistry*, 2014. **86**(10): 5131–5135.
36. Bergman, H.-M. et al., Quantitative mass spectrometry imaging of small-molecule neurotransmitters in rat brain tissue sections using nanospray desorption electrospray ionization. *Analyst*, 2016. **141**(12): 3686–3695.
37. Porta, T. et al., Quantification in MALDI-MS imaging: What can we learn from MALDI-selected reaction monitoring and what can we expect for imaging? *Analytical and Bioanalytical Chemistry*, 2015. **407**(8): 2177–2187.
38. Reich, R.F. et al., Quantitative MALDI-MS^n analysis of cocaine in the autopsied brain of a human cocaine user employing a wide isolation window and internal standards. *Journal of the American Society for Mass Spectrometry*, 2010. **21**(4): 564–571.
39. Rosen, E.P. et al., Influence of desorption conditions on analyte sensitivity and internal energy in discrete tissue or whole body imaging by IR-MALDESI. *Journal of the American Society for Mass Spectrometry*, 2015. **26**(6): 899–910.
40. Sugiyama, E. et al., Ammonium sulfate improves detection of hydrophilic quaternary ammonium compounds through decreased ion suppression in matrix-assisted laser desorption/ionization imaging mass spectrometry. *Analytical Chemistry*, 2015. **87**(22): 11176–11181.
41. Tomlinson, L. et al., Using a single, high mass resolution mass spectrometry platform to investigate ion suppression effects observed during tissue imaging. *Rapid Communications in Mass Spectrometry*, 2014. **28**(9): 995–1003.
42. Lanekoff, I. et al., Matrix effects in biological mass spectrometry imaging: Identification and compensation. *Analyst*, 2014. **139**(14): 3528–3532.
43. Lanekoff, I., M. Thomas, and J. Laskin, Shotgun approach for quantitative imaging of phospholipids using nanospray desorption electrospray ionization mass spectrometry. *Analytical Chemistry*, 2014. **86**(3): 1872–1880.
44. Prideaux, B. et al., High-sensitivity MALDI-MRM-MS imaging of moxifloxacin distribution in tuberculosis-infected rabbit lungs and granulomatous lesions. *Analytical Chemistry*, 2011. **83**(6): 2112–2118.
45. Stoeckli, M., D. Staab, and A. Schweitzer, Compound and metabolite distribution measured by MALDI mass spectrometric imaging in whole-body tissue sections. *International Journal of Mass Spectrometry*, 2007. **260**(2–3): 195–202.
46. Hochart, G., G. Hamm, and J. Stauber, Label-free MS imaging from drug discovery to preclinical development. *Bioanalysis*, 2014. **6**(20): 2775–2788.
47. Prideaux, B. et al., Mass spectrometry imaging of levofloxacin distribution in TB-infected pulmonary lesions by MALDI-MSI and continuous liquid microjunction surface sampling. *International Journal of Mass Spectrometry*, 2015. **377**: 699–708.
48. Yunsheng, H. et al., Visualization of first-pass drug metabolism of terfenadine by MALDI-imaging mass spectrometry. *Drug Metabolism Letters*, 2008. **2**(1): 1–4.

49. Kallback, P. et al., Novel mass spectrometry imaging software assisting labeled normalization and quantitation of drugs and neuropeptides directly in tissue sections. *Journal of Proteomics*, 2012. **75**(16): 4941–4951.

50. Lagarrigue, M. et al., Localization and in situ absolute quantification of chlordecone in the mouse liver by MALDI imaging. *Analytical Chemistry*, 2014. **86**(12): 5775–5783.

51. Vismeh, R. et al., Localization and quantification of drugs in animal tissues by use of desorption electrospray ionization mass spectrometry imaging. *Analytical Chemistry*, 2012. **84**(12): 5439–5445.

52. Parson, W.B. et al., Analysis of chloroquine and metabolites directly from whole-body animal tissue sections by liquid extraction surface analysis (LESA) and tandem mass spectrometry. *Journal of Mass Spectrometry*, 2012. **47**: 1420.

53. Pirman, D.A. et al., Identifying tissue-specific signal variation in MALDI mass spectrometric imaging by use of an internal standard. *Analytical Chemistry*, 2013. **85**(2): 1090–1096.

54. Lanekoff, I., M. Thomas, and J. Laskin, Shotgun approach for quantitative imaging of phospholipids using nanospray desorption electrospray ionization mass spectrometry. *Analytical Chemistry*, 2014. **86**(3): 1872–1880.

55. Hankin, J.A. and R.C. Murphy, Relationship between MALDI IMS intensity and measured quantity of selected phospholipids in rat brain sections. *Analytical Chemistry*, 2010. **82**(20): 8476–8484.

56. Cerruti, C.D. et al., *Analytical and Bioanalytical Chemistry*, 2011. **401**: 75.

57. Wang, H.-Y.J. et al., MALDI-mass spectrometry imaging of desalted rat brain sections reveals ischemia-mediated changes of lipids. *Analytical and Bioanalytical Chemistry*, 2012. **404**(1): 113–124.

58. Wang, H.-Y.J., C.B. Liu, and H.-W. Wu, A simple desalting method for direct MALDI mass spectrometry profiling of tissue lipids. *Journal of Lipid Research*, 2011. **52**(4): 840–849.

59. Goodwin, R.J.A. et al., A solvent-free matrix application method for matrix-assisted laser desorption/ionization imaging of small molecules. *Rapid Communications Mass Spectrometry*, 2010. **24**: 1682.

60. Sun, N. and A. Walch, Qualitative and quantitative mass spectrometry imaging of drugs and metabolites in tissue at therapeutic levels. *Histochemistry and Cell Biology*, 2013. **140**(2): 93–104.

61. Goodwin, R.J.A. et al., Use of a solvent-free dry matrix coating for quantitative matrix-assisted laser desorption ionization imaging f 4-bromophenyl-1,4-diazabicyclo(3.2.2)nonane-4-carboxylate in rat brain and quantitative analysis of the drug from laser microdissected tissue regions. *Analytical Chemistry*, 2010. **82**: 3868.

62. Nilsson, A. et al., Mass spectrometry imaging in drug development. *Analytical Chemistry*, 2015. **87**(3): 1437–1455.

63. Greer, T., R. Sturm, and L. Li, Mass spectrometry imaging for drugs and metabolites. *Journal of Proteomics*, 2011. **74**(12): 2617–2631.

64. Marsching, C. et al., Quantitative imaging mass spectrometry of renal sulfatides: Validation by classical mass spectrometric methods. *Journal of Lipid Research*, 2014. **55**(11): 2343–2353.

65. Reyzer, M.L. et al., Direct analysis of drug candidates in tissue by matrix-assisted laser desorption/ionization mass spectrometry. *Journal of Mass Spectrometry*, 2003. **38**: 1081.

66. Nilsson, A. et al., Fine mapping the spatial distribution and concentration of unlabeled drugs within tissue micro-compartments using imaging mass spectrometry. *PLoS ONE*, 2010. **5**(7): e11411.

67. Groseclose, M.R. and S. Castellino, A mimetic tissue model for the quantification of drug distributions by MALDI imaging mass spectrometry. *Analytical Chemistry*, 2013. **85**(21): 10099–10106.
68. Takai, N., Y. Tanaka, and H. Saji, Quantification of small molecule drugs in biological tissue sections by imaging mass spectrometry using surrogate tissue-based calibration standards. *Mass Spectrometry (Tokyo, Japan)*, 2014. **3**(1): A0025.
69. Schulz, S. et al., DMSO-enhanced MALDI MS imaging with normalization against a deuterated standard for relative quantification of dasatinib in serial mouse pharmacology studies. *Analytical and Bioanalytical Chemistry*, 2013. **405**(29): 9467–9476.
70. Pirman, D.A. et al., Quantitative MALDI tandem mass spectrometric imaging of cocaine from brain tissue with a deuterated internal standard. *Analytical Chemistry*, 2013. **85**(2): 1081–1089.
71. Bokhart, M.T. et al., Quantitative mass spectrometry imaging of emtricitabine in cervical tissue model using infrared matrix-assisted laser desorption electrospray ionization. *Analytical and Bioanalytical Chemistry*, 2015. **407**(8): 2073–2084.
72. Porta, T. et al., Single hair cocaine consumption monitoring by mass spectrometric imaging. *Analytical Chemistry*, 2011. **83**(11): 4266–4272.
73. Jadoul, L. et al., A spiked tissue-based approach for quantification of phosphatidylcholines in brain section by MALDI mass spectrometry imaging. *Analytical and Bioanalytical Chemistry*, 2015. **407**(8): 2095–2106.
74. Koeniger, S.L. et al., A quantitation method for mass spectrometry imaging. *Rapid Communications in Mass Spectrometry*, 2011. **25**(4): 503–510.
75. Takai, N. et al., Quantitative analysis of pharmaceutical drug distribution in multiple organs by imaging mass spectrometry. *Rapid Communications in Mass Spectrometry*, 2012. **26**(13): 1549–1556.
76. Bocklitz, T.W. et al., Deeper understanding of biological tissue: Quantitative correlation of MALDI-TOF and Raman imaging. *Analytical Chemistry*, 2013. **85**(22): 10829–10834.
77. Forest, S. et al., Surface plasmon resonance imaging-MALDI-TOF imaging mass spectrometry of thin tissue sections. *Analytical Chemistry*, 2016. **88**(4): 2072–2079.
78. Aichler, M. et al., Spatially resolved quantification of gadolinium(III)-based magnetic resonance agents in tissue by MALDI imaging mass spectrometry after in vivo MRI. *Angewandte Chemie-International Edition*, 2015. **54**(14): 4279–4283.
79. Goodwin, R.J.A. et al., Qualitative and quantitative MALDI imaging of the positron emission tomography ligands raclopride (a D2 dopamine antagonist) and SCH 23390 (a D1 dopamine antagonist) in rat brain tissue sections using a solvent-free dry matrix application method. *Analytical Chemistry*, 2011. **83**(24): 9694–9701.
80. Drexler, D.M. et al., Utility of quantitative whole-body autoradiography (QWBA) and imaging mass spectrometry (IMS) by matrix-assisted laser desorption/ionization (MALDI) in the assessment of ocular distribution of drugs. *Journal of Pharmacological and Toxicological Methods*, 2011. **63**(2): 205–208.
81. Solon, E.G. et al., Autoradiography, MALDI-MS, and SIMS-MS imaging in pharmaceutical discovery and development. *AAPS Journal*, 2010. **12**: 11.
82. Schadt, S. et al., Investigation of figopitant and its metabolites in rat tissue by combining whole-body autoradiography with liquid extraction surface analysis mass spectrometry. *Drug Metabolism and Disposition*, 2012. **40**(3): 419–425.
83. Goodwin, R.J.A., Sample preparation for mass spectrometry imaging: Small mistakes can lead to big consequences. *Journal of Proteomics*, 2012. **75**(16): 4893–4911.
84. Hamm, G. et al., *Journal of Proteomics*, 2012. **75**: 4952.
85. Landgraf, R.R. et al., Considerations for quantification of lipids in nerve tissue using matrix-assisted laser desorption/ionization mass spectrometric imaging. *Rapid Communications in Mass Spectrometry*, 2011. **25**(20): 3178–3184.

86. Takai, N. et al., Quantitative imaging of a therapeutic peptide in biological tissue sections by MALDI MS. *Bioanalysis*, 2013. **5**(5): 603–612.
87. Hsieh, Y. et al., Matrix-assisted laser desorption/ionization imaging mass spectrometry for direct measurement of clozapine in rat brain tissue. *Rapid Communications in Mass Spectrometry*, 2006. **20**(6): 965–972.
88. Pirman, D.A. and R.A. Yost, Quantitative tandem mass spectrometric imaging of endogenous acetyl-L-carnitine from piglet brain tissue using an internal standard. *Analytical Chemistry*, 2011. **83**(22): 8575–8581.
89. Liu, X. and A.B. Hummon, Mass spectrometry imaging of therapeutics from animal models to three-dimensional cell cultures. *Analytical Chemistry*, 2015. **87**(19): 9508–9519.
90. Shroff, R. et al., Quantification of plant surface metabolites by matrix-assisted laser desorption-ionization mass spectrometry imaging: Glucosinolates on Arabidopsis thaliana leaves. *Plant Journal*, 2015. **81**(6): 961–972.
91. Bunch, J., M.R. Clench, and D.S. Richards, *Rapid Communications in Mass Spectrometry*, 2004. **18**: 3051.
92. Clemis, E.J. et al., Quantitation of spatially-localized proteins in tissue samples using MALDI-MRM imaging. *Analytical Chemistry*, 2012. **84**(8): 3514–3522.
93. Lee, Y.J. et al., Use of mass spectrometry for imaging metabolites in plants. *Plant Journal*, 2012. **70**(1): 81–95.
94. Almeida, R. et al., Quantitative spatial analysis of the mouse brain lipidome by pressurized liquid extraction surface analysis. *Analytical Chemistry*, 2015. **87**(3): 1749–1756.
95. Mandal, M.K. et al., Development of sheath-flow probe electrospray ionization mass spectrometry and its application to real time pesticide analysis. *Journal of Agricultural and Food Chemistry*, 2013. **61**(33): 7889–7895.
96. Gurdak, E. et al., VAMAS interlaboratory study for desorption electrospray ionization mass spectrometry (DESI MS) intensity repeatability and constancy. *Analytical Chemistry*, 2014. **86**(19): 9603–9611.
97. Önnerfjord, P. et al., Homogeneous sample preparation for automated high throughput analysis with matrix-assisted laser desorption/ionisation time-of-flight mass spectrometry. *Rapid Communications in Mass Spectrometry*, 1999. **13**(5): 315–322.
98. Gessel, M.M., J.L. Norris, and R.M. Caprioli, MALDI imaging mass spectrometry: Spatial molecular analysis to enable a new age of discovery. *Journal of Proteomics*, 2014. **107**: 71–82.
99. Hankin, J.A., R.M. Barkley, and R.C. Murphy, Sublimation as a method of matrix application for mass spectrometric imaging. *Journal of the American Society for Mass Spectrometry*, 2007. **18**(9): 1646–1652.
100. Yang, J. and R.M. Caprioli, Matrix sublimation/recrystallization for imaging proteins by mass spectrometry at high spatial resolution. *Analytical Chemistry*, 2011. **83**(14): 5728–5734.
101. Robichaud, G. et al., Infrared matrix-assisted laser desorption electrospray ionization (IR-MALDESI) imaging source coupled to a FT-ICR mass spectrometer. *Journal of the American Society for Mass Spectrometry*, 2012. **24**(1): 92–100.

3 Long-Range Molecular Interactions Involved in the Retention Mechanisms of Liquid Chromatography

*Victor David, Nelu Grinberg, and
Serban C. Moldoveanu*

CONTENTS

3.1   BRIEF INTRODUCTION

Several fundamental driving forces form the basis of intermolecular and intramolecular interactions in chemical and biochemical systems. Among these are interactions such as hydrophobic, London dispersion, hydrogen bonding, and electrostatic forces. In the past three decades, the sophistication and power of techniques to investigate these processes have developed at an unprecedented rate [1]. The mass transfer of species between two phases is a consequence of differences between interactions of transferred species with phase components. Chromatographic process as a distribution between two phases is a reversible transfer of species between a mobile phase and a stationary phase, and it basically relies on differences between chemical affinities based on interactions of solutes taking part to this process. As pointed out in a recent paper, chromatography is an excellent analytical method to study molecular interactions [2]. In many chromatographic types, and generally in separation science, some of these interactions are long range, which does not support chemical bonds or permanent fixation of certain species on stationary phase. Others are ionic bonds, which also play an important role in ion chromatography and in hydrophilic interaction liquid chromatography (HILIC). These bonds are not long range, but they can be involved in fast equilibria, allowing the exchange process to take place. In liquid chromatography (LC), various types of bonds are involved in the interactions between mobile phase components and analyte molecule, between stationary phase and analyte molecule, as well as between mobile phase components and the surface of stationary phase. From this point of view, the interaction process in LC is much more complicated compared to gas chromatography (GC), where at least the interactions of molecules of analytes and stationary phase surface with mobile phase components are almost negligible [3]. This allows simple classification schemes for stationary phases in GC, which are based on intermolecular interactions between analyte molecule and functionalities from the surface of stationary phase [4].

   Interaction between two molecules is basically found in Schrodinger equation ($\hat{H}\Psi = E\Psi$), when the Hamiltonian operator $\hat{H}$ acting on wavefunction Ψ generates the energetic eigenvalue of the system, E. When two molecules, denoted by A and B, interact, then their system is described by the Hamiltonian $\hat{H}_{AB} = \hat{H}_A + \hat{H}_B + \hat{V}_{AB}$, where the operator $\hat{V}_{AB}$ represents the interaction potential between A and B. However, this equation cannot be solved exactly and the main formalism is comprised in the perturbation theory, when $\hat{V}_{AB}$ is given various analytical expressions depending on the components under interaction [5]. Information on the importance of intermolecular potential and the semiclassical perturbation theory treatment of short- and long-range forces can be found in recent specialized books, with considerable advances in quantum theory of intermolecular forces and *ab initio* methods for calculating the interaction energy [6,7].

3.2 LONG-RANGE VERSUS SHORT-RANGE INTERACTIONS

Intermolecular forces can be attractive or repulsive. Generally, they are separated into two main groups, depending on the distance between interacting species, namely the short-range and long-range forces. Short-range forces occur when the centers of the molecules are separated by less than 2 Å. Short-range forces tend to be repulsive, where the long-range forces that act outside the 3 Å range are attractive. Long-range forces include van der Waals forces and other interactions such as hydrogen bonding (not usually included in van der Waals type). Intermolecular forces are responsible for differences between actual behavior of gases and that predicted by the ideal gas law. They are responsible for most properties of all liquids, such as viscosity, diffusion, and surface tension.

Moreover, the dissolution of gases, liquids, or solids in various solvents depends on intermolecular forces, and the *like-to-like* principle of explaining solubilization of solutes in a liquid solvent is based on similar interactions between solute molecules and solvating molecules as those occurring between solvent molecules. Not only with liquids, but the role played by intermolecular forces in terms of electron density is fundamental to the understanding, for example, the self-assembly process of molecules in inorganic dimers and the formation of a molecular crystal [8].

Different models describing molecular interactions differentiate them based on their energy. This difference consists of interactions with energy lower than 40 kJ/mol and higher than 40 kJ/mol [9]. The relation between the hydrophobic forces acting between hydrophobic solute molecules and macroscopic hydrophobic surfaces has been a topic of considerable interest from theoretical point of view. Recent improved surface preparation techniques, together with combination between surface force measurements and atomic force microscopy imaging have made it possible to explain the long-range part of this interaction (at distances >200 Å) that can be observed between certain surfaces [10].

3.3   GENERAL MODEL FOR THE ELECTROSTATIC INTERACTIONS BETWEEN TWO CHARGE COLLECTIONS

The simplest model considers a molecule as a charge collection, with positive and negative centers. These charges interact with the surrounding medium by electrostatic forces, and for modeling this process the molecule is represented as charge collection as depicted in Figure 3.1. This model can explain the interactions in ion chromatography and also part of the behavior of polar compounds when they are studied by HPLC with polar stationary phase, such as those containing halogen atoms bound to a phenylsilica-based stationary phase [11]. Recent computing programs allow simulations of charge distribution for stationary phase or analytes participating to the elution process, such as the examples given in Figure 3.2 that show different charges on ethylphenyl and cyanopropyl bonded phases on a silica surface computed by means of MarvinSketch program [12]. In Figure 3.3, this possibility was checked for polar compounds with different polarities, separated by means of hydrophilic interaction-based liquid chromatography [13].

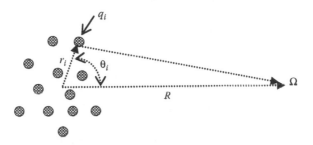

FIGURE 3.1   Representation of a molecule as a collection of charges q_i and the corresponding electric field created in a point Ω situated outside the molecule.

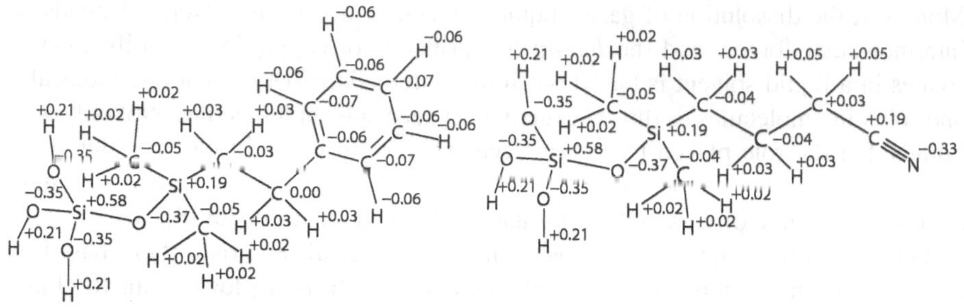

FIGURE 3.2   Charge distributions on model molecules simulating ethylphenyl and cyano-propyl bonded phases on a silica surface (with MarvinSketch program).

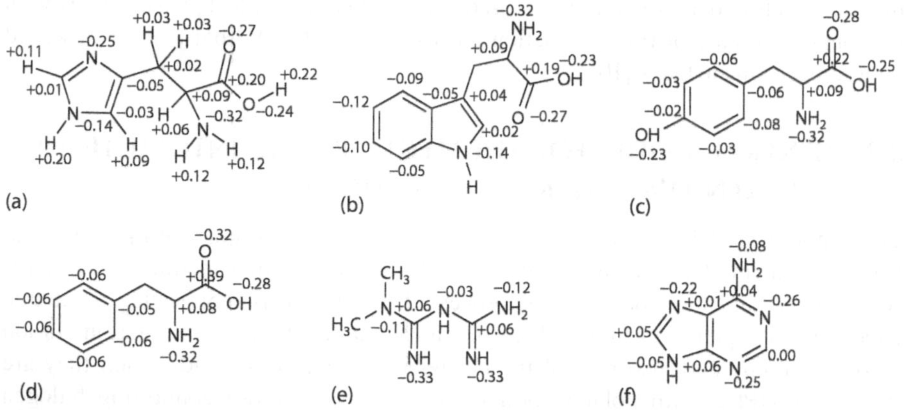

FIGURE 3.3   Charge distributions calculated by means of MarvinSketch program for six polar molecules participating to the chromatographic elution under Zic-Hilic mechanism. (a) Histidine, (b) tryptophan, (c) tyrosine, (d) phenylalanine, (e) metformin, and (f) adenine.

According to this model, each q_i charge of a molecule is characterized by a position denoted by r_i from the center of the molecule. The electric potential $V(R)$ created by this charge distribution at a distance R from the center of the charge distribution can be written as vectorial expression as following:

$$V(R) = \sum_i V_i = \sum_i \frac{q_i}{\left|\vec{R} - \vec{r_i}\right|} \tag{3.1}$$

Here, the vectors involved in this model have the following expressions:

$$\vec{r_i} = (x_i, y_i, z_i) = x_i\vec{i} + y_i\vec{j} + z_i\vec{k} \tag{3.2}$$

$$\vec{R} - \vec{r_i} = (X - x_i)\vec{i} + (Y - y_i)\vec{j} + (Z - z_i)\vec{k} \tag{3.3}$$

$$\left|\vec{R} - \vec{r_i}\right| = [(X - x_i)^2 + (Y - y_i)^2 + (Z - z_i)^2]^{1/2} \tag{3.4}$$

Taylor series of the potential $V(R)$ from Equation 3.1 leads to a complicated expression of V as depending on the charges q_i and their positions in molecule, in the following form [14]:

$$V(R) = \frac{1}{R}\sum_i q_i + \frac{1}{R^2}\left[\frac{X}{R}\sum_i q_i x_i + \frac{Y}{R}\sum_i q_i y_i + \frac{Z}{R}\sum_i q_i z_i\right]$$

$$+\frac{1}{R^3}\left[\frac{1}{2}\left(\frac{3X^2}{R^2}-1\right)\sum_i q_i x_i^2 + \frac{1}{2}\left(\frac{3Y^2}{R^2}-1\right)\sum_i q_i y_i^2 + \frac{1}{2}\left(\frac{3Z^2}{R^2}-1\right)\sum_i q_i z_i^2 \right. \quad (3.5)$$

$$\left.+\frac{3XY}{R^2}\sum_i q_i x_i y_i + \frac{3XZ}{R^2}\sum_i q_i x_i z_i + \frac{3YZ}{R^2}\sum_i q_i y_i z_i\right] + \frac{1}{R^4}[\text{upper terms}]$$

The net charge of the distribution (named as monopole) is denoted by q:

$$q = \sum_i q_i \qquad (3.6)$$

Polarity of a molecule that has the meaning of separation between center of positive charge from that of negative charge is described by the dipole moment, which is denoted by μ and given by the sum:

$$\mu = \sum_i q_i r_i \qquad (3.7)$$

The quadrupole moment Q is defined by the expression:

$$Q = \sum_i q_i r_i r_i \qquad (3.8)$$

The potential given by Equation 3.5 can be rewritten in a dependence of Legendre polynomials P_k and the angles θ_i between r_i and R [15]:

$$V(R) = \sum_{k=0}^{\infty} \frac{1}{R^{k+1}} \sum_i q_i r_i^k P_k(\cos\theta_i) \qquad (3.9)$$

When two charge distributions interact, $\{q_i; r_i\}$ and $\{q_j; r_j\}$, separated by a distance $\vec{R}$ between their mass centers, written as a vector with coordinates X, Y, Z, their interaction energy $E_{\text{interaction}}$ is given by the equation:

$$E_{\text{interaction}} = \sum_j q_j V(\vec{R}) \qquad (3.10)$$

Using a Taylor series for this energy, the expression of $E_{\text{interaction}}$ becomes:

$$E_{\text{interaction}} = \sum_j q_j V(\vec{R}) + \sum_j q_j\left(x_j\frac{\partial V}{\partial X} + y_j\frac{\partial V}{\partial Y} + z_j\frac{\partial V}{\partial Z}\right)$$

$$+\frac{1}{2!}\sum_j\left[q_j\left(x_j^2\frac{\partial^2 V}{\partial^2 X^2} + ...\right) + 2y_i z_j\frac{\partial^2 V}{\partial Y\partial Z} + ...\right] \qquad (3.11)$$

Using this equation, we can obtain the energy of different types of interactions, such as ion–ion, ion–dipole, ion–quadrupole, dipole–dipole, dipole–quadrupole, and so on. This task will be avoided in this description, and only the main important interactions will be further discussed briefly and analyzed from the point of view of the liquid chromatographic process.

Interaction between two ions (charge to charge interaction) can be described by the Coulomb's law, which is the simplest interaction model involving two point charges situated at a distance d in a continuum solvating medium characterized by a dielectric constant ε relative to vacuum (ε for vacuum is taken as 1):

$$E_{\text{ion–ion}} = \frac{q_1 q_2}{4\pi\varepsilon\varepsilon_0 d} \tag{3.12}$$

In this equation, ε_0 represents the vacuum permittivity, that is $8.85 \cdot 10^{-12}$ CV^{-1} m^{-1}. The free energy E can be written for two ions whose point charges are $q_1 = z_1 e$ and $q_2 = z_2 e$, respectively, where the value of the elementary charge e is 1.602×10^{-12} C, as following:

$$E_{\text{ion–ion}} = \frac{z_1 z_2 e^2}{4\pi\varepsilon\varepsilon_0 d} \tag{3.13}$$

Depending on the signs of the two interacting charges, $E_{\text{ion–ion}}$ can be positive, that is, the charges have the same sign, and negative sign for the charge of opposite signs.

For the ion–dipole interaction ($E_{\text{ion–dipole}}$), between a charge $q = ze$ and a dipole moment μ separated by a distance r, the energy will also be associated with the entropy of the system, and in a good approximation this is given by the expression:

$$E_{\text{ion–dipole}} = \frac{(ze)^2 \mu^2}{3k_B T (4\pi\varepsilon_0\varepsilon)^2 r^2} \tag{3.14}$$

where:
T represents the absolute temperature
k_B is the Boltzmann constant

The ion–dipole interaction is weaker than ion–ion, but when the distance r is shorter than 2–4 Å, it can be significantly higher than $k_B T$, which is about 4×10^{-21} at 300 K [16]. An important parameter in the treatment of intermolecular forces is dielectric constant, which is a bulk constant. For this reason, the values of several very common solvents utilized in liquid chromatography, relative to vacuum, are following: water—77.46; methanol—32.7; acetonitrile—37.5; acetone—20.7; ethanol—24.6; isopropanol—19.9; acetic acid—6.2; dioxane—2.21.

Interaction between two dipole moments is much more complicated due to geometry involved in such process, besides the fact that the interaction is temperature dependent [16]. Similar to the interaction between an ion and a dipole, the interaction energy ($E_{\text{dipole–dipole}}$) between two dipoles μ_1 and μ_2 separated by a distance r, known also as Keesom energy, has the following expression:

$$E_{\text{dipole–dipole}} = -\frac{2\mu_1^2\mu_2^2}{3k_BT(4\pi\varepsilon_0\varepsilon)^2r^6} \tag{3.15}$$

Induced dipole moment (denoted by μ_{ind}) in a nonpolar molecule occurs when an electric field is applied due to the presence of neighboring polar or ionic species. By this perturbational electric field, the electron clouds of nonpolar molecule can be distorted, and the molecule becomes polar. This is a vector and it is proportional to the intensity of external electric field acting on the molecule, and the constant α_{ind} linking the value of μ_{ind} and intensity of electric field $\vec{V}_e$ represents the polarizability of the molecule and is measured in $Cm^2\ V^{-1}$:

$$\vec{\mu}_{\text{ind}} = \alpha_{\text{ind}}\vec{V} \tag{3.16}$$

Molecular polarizability plays an important role in studying the interaction systems between polar and nonpolar species, and for this purpose an important literature is focused on this molecular parameter [17–19]. The molecular polarizability can be calculated from Clausius–Mossoti expression [20]:

$$\alpha_{\text{ind}} = \left(\frac{n^2-1}{n^2+2}\right) \cdot \frac{3M\varepsilon_0}{\rho_{\text{solv}}N} + \frac{\mu^2}{3k_BT} \tag{3.17}$$

where:

n is the refractive index of the bulk solvent

M is the molecular weight

ρ_{solv} is the density

The other parameters are the same as previously defined in Moldoveanu and Savin [19].

The literature reports numerous studies proposing different models for studying the influence of a solvent on the molecular interactions [21–24].

The free energy of interaction between a dipole μ_1 and an induced dipole in molecule 2 due to the polarizability α_2 is given by the equation:

$$E_{\text{dip–ind}} = -\frac{\mu_1^2\alpha_2}{(4\pi\varepsilon_0\varepsilon)^2r^6} \tag{3.18}$$

Thus, the general expression for calculating the average energy of interaction between two molecules with dipole moments μ_1 and μ_2, and polarizabilities α_1 and α_2, respectively, which is also known as Debye interaction energy becomes:

$$E_{\text{dip–ind}} = -\frac{\mu_1^2\alpha_2 + \mu_2^2\alpha_1}{(4\pi\varepsilon_0\varepsilon)^2r^6} \tag{3.19}$$

Theoretically, the strength of interaction is given by the following order: $E_{\text{ion–ion}} > E_{\text{ion–dipole}} > E_{\text{dipole–dipole}} > E_{\text{dip–ind}} > E_{\text{dipole–quadrupole}}$. In reality, they have a wide interval

TABLE 3.1

Usual Values for the Energy of Different Types of Intermolecular Forces in Comparison with Interatomic Forces

Interaction Type	Energy in kJ/mol	Observations
Covalent	200–400	This interaction is not an intermolecular but an energetic reference for the other forces
Ion–ion	200–800	Involved in forming the lattice of a crystal
Atom–atom intermetallic	100–350	Involved in forming the metallic lattice
Electron pair donor–electron pair acceptor (EPD–EPA)	10–200	Also referred to as charge transfer interactions, and it is considered as a chemical bond
Ion–dipole	60–130	In RP-LC: molecular dipole of solute and metallic ions from surface of stationary phase
Hydrogen bonding	5–40	In all mechanisms where silanol is present on surface of stationary phase
Dipole–dipole	20–40	Interfering in RP-LC mechanism when stationary phase contains polar embedded groups, such as urea, or carbamate
Dipole–induced dipole	4–20	It depends on the presence of a polar molecule in a nonpolar solvating environment
Dispersion	8–30	Cumulative and consequently these forces could be very energetic
π–π interactions	40–80	Between aromatic and unsaturated structures, but they can also occur between anions or cations with π systems

of values, and for some particular systems, this strength can deviate from the theoretical rule. The range of variation of these energies, in comparison with energies characterizing the chemical bonds, or other interactions that will be discussed further are given in Table 3.1 [25].

Molecular structure is also an important factor for interactions between a solute molecule and its surrounding molecules or a surface nearby. It may depend on several external (such as temperature, solvent, pressure) and internal factors (torsion angle, internal energy). Overall, molecular geometry is determined by internal energy and interactions with surrounding molecules. That means that the conformation of an isolated molecule may differ to the conformation in a liquid medium, subjected to high pressure and many interactions.

3.4   ION ENERGY IN A SOLVATING MEDIUM

The partition of ionic species between mobile and stationary phase is important in at least two widely used LC separation mechanisms, such as ion chromatography and hydrophilic interaction LC. This partition can be thermodynamically modeled by difference of the ion energy in the two phases. For this purpose, it is necessary to have a relationship that describes the interaction energy of an ionic species with a surrounding medium, characterized by dielectric constant ε_{solv}. In order to calculate this energy, it is necessary to assume that in the lack of any interactions with other ionic species, an ion has its own free energy, which is equal to the electrostatic work done for forming that ion in the presence of solvating medium considered as a continuum one and characterized by its dielectric constant ε_{solv}. For an ion with the charge q and radius r, the increase of the charge with dq, in vacuum and in the medium characterized by dielectric constant ε_{solv}, will require the following energies:

$$(dE)_{vacuum} = \frac{qdq}{4\pi\varepsilon_0 r} \tag{3.20}$$

$$(dE)_{solv} = \frac{qdq}{4\pi\varepsilon_0\varepsilon_{solv} r} \tag{3.21}$$

The total free energies for charging one ion to the final charge ze in vacuum and in solvating medium will be obtained by integrating this energy over the interval $[0; ze]$, for both energies:

$$(E_{ion})_{vacuum} = \frac{1}{4\pi\varepsilon_0 r}\int_0^{ze} qdq = \frac{z^2 e^2}{8\pi\varepsilon_0 r} \tag{3.22}$$

$$(E_{ion})_{solv} = \frac{1}{4\pi\varepsilon_0\varepsilon r}\int_0^{ze} qdq = \frac{z^2 e^2}{8\pi\varepsilon_0\varepsilon_{solv} r} \tag{3.23}$$

The sign is positive for each energy, due to the fact that the formation of ion requires an external work equal to this energy. For calculating the energy of a mole of ions, the expression (3.23) is multiplied by Avogadro's number, N. The difference of the two energy, $(E_{ion})_{solv} - (E_{ion})_{vacuum}$, gives the energy necessary to transfer the ion at constant volume from vacuum in solvating medium, becoming:

$$(\Delta E_{ion})_{solv} = N\frac{z^2 e^2}{8\pi\varepsilon_0 r}\left(\frac{1}{\varepsilon_{solv}} - 1\right) \tag{3.24}$$

In literature, this energy is known as Born energy, and its sign is negative because $\varepsilon_{solv} > 1$. Negative value indicates that the event is thermodynamically favorable for the transfer of ion from vacuum into the solvating medium. The negative value of $(\Delta E_{ion})_{solv}$ also indicates that the solubility of an ionic species in a solvent increases with the increase in the solvent dielectric constant. However, the Born model to calculate the solvation energy of ions is a simplifying one because it does not take into consideration the energetic contribution necessary to create the cavity in the solvent for placing the ion to the entire process energy. Although this model represents only a rough approximation of complex process of dissolution, a fair linear dependence having negative slope was noticed between the solubility values of monovalent ions and $1/\varepsilon$ of the various solvents [26].

This procedure can be continued for modeling the transfer of an ion between two liquid phases, when solv $= \alpha$ and β. The difference of energies at the ion transfer from phase α to the phase β, denoted by $\Delta(\Delta E_{ion})$, becomes:

$$\Delta(\Delta E_{ion}) - (\Delta E_{ion})_\beta \quad (\Delta E_{ion})_\alpha - N \frac{z^2 e^2}{8\pi\varepsilon_0 r}\left(\frac{1}{\varepsilon_\beta} - \frac{1}{\varepsilon_\alpha}\right) \tag{3.25}$$

This difference is negative such that the process to be thermodynamically favorable to the transfer of ion from phase α to phase β, when $\varepsilon_\beta > \varepsilon_\alpha$. This means that the ions are easily transferred from a polar phase to a more polar phase.

3.5 OTHER IMPORTANT INTERMOLECULAR INTERACTIONS

Two intermolecular interactions that cannot be comprised in a mathematical formalism are hydrogen bonding and π–π interaction (stacking). They play an important role in many retention mechanisms in LC. Hydrogen bonding (or H-bonding) is present between a polar hydrogen and an electronegative atom (from another molecule, or from a different part of the same molecule). The other interaction occurs when two aromatic systems are close, but this type of interaction may occur between anions or cations with π systems.

According to their strength of formation, H-bonds are classified into three main categories: weak characterized by low molar energy ($\Delta E < 20$ kJ/mol), intermediate (or medium-strength) with energy between 20 and 40 kJ/mol, and strong H-bonds with energy higher than 40 kJ/mol [27]. The enthalpy of hydrogen bonding depends on the nature of the atoms to which the hydrogen is bound and also of the acceptor atom. For example, the average energy of a hydrogen bond from an OH group is between 20–25 kJ/mol and 8–12 kJ/mol for NH_2 groups, but it is specific to the interacting molecules (O–H$\cdots$N with about 29 kJ/mol, O–H$\cdots$O with about 21 kJ/mol, N–H$\cdots$N with about 13 kJ/mol, N–H$\cdots$O with about 8 kJ/mol) and depends on temperature [28].

In liquid chromatography, H-bonds are typically formed in mobile phase, but this type of interaction also occurs between various functional groups of analytes containing H-donor or acceptor and residual silanol groups present on the surface of stationary phase. The interference of this interaction in the retention mechanism, for example in reversed-phase liquid chromatography (RP-LC), depends on the density of silanol groups from stationary phase. There is a wide variety of available RP-LC stationary phases, which differ on the density of hydrocarbon ligand bonded on silica surface, and consequently on the density of residual silanol groups. These groups are also able to take part to ion exchange-like processes, or to interact with functional groups from solute molecule by means of the mentioned H-bonds. The literature proposes many empirical tests, which refer to the silanol activity, but only a few focused on H-bonds. For example, Tanaka's test uses a mobile phase composed of methanol/water (30:70, v/v); a column temperature of 40°C; and test compounds uracil (t_0), caffeine, and phenol [29]. Hydrogen-bonding capacity is proposed to be the ratio $k_{caffeine}/k_{phenol}$. Engelhardt's test suggests using a mobile phase containing methanol/water (49:51, or 55:45, v/v)

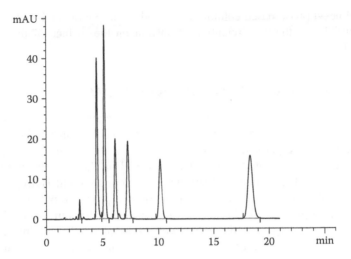

FIGURE 3.4 HPLC-DAD chromatogram of a solution containing six benzodiazepines (elution order: bromazepam, nitrazepam, flunitrazepam, alprazolam, diazepam, and medazepam). (Experimental conditions: Zorbax XDB Phenyl (125 mm × 4.6 mm × 5 μm d.p.); flow-rate: 1 mL/min; column temperature: 25°C; mobile phase composition: 65% MeOH and 35% phosphate buffer with pH = 7.)

and *p*-ethylaniline as test analyte [30]. In this case, the asymmetry of test compound measured at 5% of peak height indicates the silanol activity.

Interaction of type π–π is taking place between solutes containing aromatic or unsaturated bonds (C=C, C≡N) and stationary phases containing aromatic moieties, but the retention mechanism is a combination of van der Waals and π–π interaction. A classic example is given by stationary phases containing phenyl rings bound to silica surface, or polymeric stationary phases based on poly(styrene-divinylbenzene). An example of separation of several benzodiazepines in reversed-phase mechanism on a phenyl silica-based column is given in Figure 3.4 [31]. More unusual stationary phases contain, for example, porphyrin derivatives or phthalocyanine derivatives, bound to aminopropyl silica gels [32], which were used in separation of polynuclear aromatic hydrocarbons (PAHs) based on π–π electron interaction between these solutes and aromatic stationary phase. In particular, the aminopropyl silica gel modified with copper phthalocyanine (Cu-PCS) showed the strongest π–π electron interaction with PAHs.

Cyanopropylsilica can also be involved in π–π interaction with aromatic or unsaturated solutes. Experimental studies performed on cyano and phenyl silica stationary phase showed that π–π interactions can contribute to retention on these columns, while dipole–dipole interactions are more likely to be significant for the retention of polar aliphatic solutes on cyano columns. Unlike the use of methanol/water as mobile phase, both π–π and dipole–dipole interactions can be suppressed for mobile phases based on acetonitrile/water [33,34]. Differences between selectivity on alkyl and π-stationary phases were assessed and then used for two-dimensional separation of natural samples [35]. Another application refers to the possibility of injecting high volume of aromatic solvents (benzene, toluene, ethylbenzene, and propylbenzene) on phenyl-hexyl phase-based column, when both hydrophobic and π–π interactions are responsible for the high retention of solvent on the surface of this stationary phase [36].

3.6 MODELS FOR SOLUTE–SOLVENT INTERACTION

An accurate model to describe solute–solvent interactions for being used to predict solution-phase properties, such as free enthalpy of solvation, solubility, partition constants, membrane permeability, is difficult to design. So far, only approximate models have been elaborated, many of them based on the concept of solute molecule as a charge distribution, and solvent as a continuum and isotropic medium. One model to describe solute–solvent interaction was proposed by Onsager, relying on the concept of reaction field that characterizes the continuum assumed solvating medium and its interaction with charged collection modeled solute molecule [37]. The interaction energy between charged collection, denoted by $\{q_k\}_{k=1,m}$, and reaction field Φ_R is called as polarization energy (E_{pol}), which was demonstrated by Kirkwood [38–40], and is given by the expression:

$$E_{pol} = \frac{1}{2}\sum_{k=1}^{m} q_k \Phi_R \tag{3.26}$$

This mathematical model, known as molecule-continuum medium, is rather complicated, and it is far to the aim of this work to describe this in detail. The expression of polarization energy is given by the formula:

$$E_{pol} = \frac{1}{2}\sum_{n=0}^{\infty} \frac{(n+1)(1-\varepsilon)}{n+(n+1)\varepsilon} \cdot \frac{Q_n}{d_m^{2n+1}} \tag{3.27}$$

where:

Q_n represents the pole of n-order of the charge collection (solute molecule)

ε is the dielectric constant of the solvating medium

d_m is the molecular diameter for the solute molecule assumed as sphere one

For example, the polarization energy for the dipole moment ($E_{pol}^{(1)}$) of the solute molecule (μ) becomes:

$$E_{pol}^{(1)} = \frac{1-\varepsilon}{1+2\varepsilon} \cdot \frac{\mu^2}{d_m^3} \tag{3.28}$$

This model was improved by Beveridge and Schnuelle, who suggested a supermolecule cluster between solute molecule and solvent molecules (so called supermolecule-continuum model), by defining two layers of solvent, with two different bulk properties: one for the inner layer of solvent molecule with thickness distanced by d_1, and dielectric constant ε_{loc}, and the other for the solvent bulk, denoted by ε [41,42]. The final expression of the polarization energy for the dipole moment ($E_{pol}^{(1)}$) is obtained in the form:

$$E_{pol}^{(1)} = \left[\frac{1-\varepsilon'_{loc}}{1+2\varepsilon_{loc}} + \frac{2(1-\varepsilon)}{1+2\varepsilon} \left(1 - \frac{1-\varepsilon'_{loc}}{1+2\varepsilon'_{loc}} \right) \right] \cdot \frac{\mu^2}{d_m^3} \tag{3.29}$$

where ε'_{loc} is given by the relation:

$$\varepsilon'_{loc} = \frac{\varepsilon_{loc}}{\varepsilon} \left[1 + \frac{(n+1)(1-\varepsilon)(1-\varepsilon_{loc})}{n+(n+1)\varepsilon} \cdot \frac{d_m^{2n+1}}{d_i^{2n+1}} \right]^{-1} \tag{3.30}$$

This model provides a general meaning of the treating solvent effects and identifies the contributions of various dielectric regions in the vicinity of the solute molecule. Meanwhile, many other studies were focused on semiempirical approaches for the general treatment of solute–solvent interactions. One such approach is based on the principle that all solution-phase processes can be modeled in terms of one or more gas-to-solution transfer processes [43]. The free energy of each gas-to-solution transfer process is calculated as the sum of the free energy of cavity formation and the free energy of solute–solvent interaction. According to this model, the free energy of cavity formation, where the solute molecule is placed, can be modeled on the basis of the total solvent-accessible surface area of the solute. The enthalpy of solute–solvent interaction is modeled on the basis of intermolecular interaction potentials calculated at many uniformly distributed points on the solvent-accessible surface of the solute [44].

Important results to estimate the solvent effects were obtained by assuming that in the evaluation of interaction energy, a solvation term proportional with the polar solvent accessible area of the molecule (SASA) can be used [45,46]. According to this theory, the effective energy of interactions for a solute "k" having n atoms is divided in two contributions: one due to the solutes interaction, E_k, and another due to the interactions of solute with the solvent molecules, E_{sol}, in a relation as following:

$$E(r) = E_k + E_{sol}(r) \tag{3.31}$$

where $r = (r_1, r_2, r_3, ..., r_n)$ represents a vector indicating the position of each atom in the solute molecule. The term representing the solute–solvent interaction $E_{solv}(r)$ was found to be related to solvent accessible area of the solute molecule (A^{SASA}) by a relation of the form:

$$E_{solv}(r) = \sum_{i=1}^{n} \sigma_i A_i^{SASA} \tag{3.32}$$

where:

σ_i is an atomic solvation parameter
A_i^{SASA} is the polar solvent accessible area of the atom i from the solute molecule

The values for A_i^{SASA} can be obtained using the following analytical expression:

$$A_i^{SASA} = S_i \prod_{j \neq i} \left[1 - \frac{p_i p_{ij} b_{ij}(r_{ij})}{S_i} \right] \tag{3.33}$$

where S_i represents the solvent accessible area of an isolated atom of radius R_i as given by

$$S_i = 4\pi(R_i + R_{\text{solv}})^2 \tag{3.34}$$

where R_{solv} is the radius of the solvent molecule. The term denoted by $b_{ij}(r_{i,j})$ represents the SASA removed from S_i due to overlapping between atoms "i" and "j" at the distance $r_{i,j}$. The calculation of $b_{ij}(r_{i,j})$ can be done using the expressions:

$$b_{ij}(r_{ij}) = 0 \quad \text{for} \quad r_{ij} > R_i + R_j + R_{\text{solv}} \tag{3.35}$$

otherwise:

$$b_{ij}(r_{ij}) = \pi(R_i + R_{\text{solv}})(R_i + R_j + 2R_{\text{solv}} - r_{ij})[1 + (R_j - R_i)r_{ij}^{-1}] \tag{3.36}$$

These values of parameters R_i, p_i, and σ_i are tabulated and can be found in literature [47]. The formulas for the calculation of SASA for organic molecules as shortly described are used in computer programs that provide values for this parameter for a variety of molecules (e.g., MarvinSketch [12]). The estimation of E_{solv} values was used with good results for the simulation of interaction of small proteins in solution and also can be used to understand the interaction of small polar molecules with a solvent [48–50]. This concept has been mentioned as important in liquid chromatography for describing quantitative dependences between the chromatographic response and SASA parameter [51]. For this reason, SASA is an important parameter in quantitative relationships for structure-retention behavior of organic compounds in LC [52], or in predicting molecular descriptor in the computer-aided drug design [53].

Solvation process is an important part of chromatographic process and can also take place between components from mobile phase and active sites from the stationary phase [54]. This process is tremendously dependent on the polarity of these sites. Thus, the polar adsorption sites are preferentially solvated by water molecules or other protic solvents, whereas the hydrophobic chains are solvated by more hydrophobic molecules, such as the organic solvents added to mobile phase [55].

3.7 ELECTROSTATIC INTERACTIONS IN HYDROPHILIC INTERACTION LIQUID CHROMATOGRAPHY AND NORMAL PHASE RETENTION MECHANISMS

Interactions between analyte and stationary phase in HILIC and normal phase (NP) mechanisms are mainly of electrostatic nature, and the principal forces involved in the partition process of analyte from mobile phase and stationary phase are ion–ion, ion–dipole, dipole–dipole, and H-bonding. Small contribution from hydrophobic interactions can also be considered. Charged analytes are generally the most suitable compounds to be separated in HILIC mechanism, because they are more hydrophilic than their uncharged forms; therefore, they are more retained on the polar or dissociated silanols from stationary phase. The highest interactions occur between charged analyte and charged sites of the stationary phase. These interactions can be attractive or repulsive, depending on the signs of charges of both the analyte and the stationary phase. Electrostatic attractions between charged analyte and opposite charged functionality from stationary phase lead to the increase of retention times, whereas electrostatic repulsions between analytes and functionalities with same charges have the opposite effect on the retention time [56].

The mobile phase in HILIC and normal-phase chromatography (NPC) is characterized by its lower polarity compared to that of the stationary phase. Typical mobile phase compositions in HILIC are made of an aqueous buffer with controlled pH and an organic solvent such as acetonitrile, or methanol, although other solvents, such as ethanol, or 2-propanol, are sometimes used as organic modifiers. The solvent strength in HILIC has the following order: water > methanol > ethanol > 2-propanol > acetonitrile > acetone > tetrahydrofuran. However, acetonitrile is the most preferred organic solvent in HILIC applications, whereas the other solvents provide insufficient analyte retention and broad or nonsymmetrical peak shapes in many separations. This refers especially to methanol, which is rarely used as the organic component in HILIC separations. The poor HILIC performance of methanol may be due to its similarity to water—both methanol and water being protic solvents. Methanol can compete to solvate the surface of silica or of other polar stationary phases used in HILIC and provide strong hydrogen bonding interactions with each other [57–59].

The solvents or additives used in NPC can be alkanes or cycloalkanes (n-pentane; n-hexane; n-heptane; i-octane; cyclopentane, cyclohexane), fluoroalkanes, chlorinated alkanes (dichloromethane; chloroform; carbon tetrachloride; propylchloride), ethers (diethyl ether; di-i-propyl ether), esters (methyl acetate; ethyl acetate), alcohols (methanol; ethanol; 1-propanol; 2-propanol), amines (pyridine; propylamine; triethylamine), and carboxylic acids or their derivatives, such as dimethylformamide [60].

HILIC represents an alternative to RP-LC separations of polar, weakly acidic or basic species, such as peptides, proteins, oligosaccharides, drugs, metabolites, and various natural compounds. Various columns can be used not only in the HILIC mechanism: bare silica gel, silica-based amino-, amido-, cyano-, carbamate-, diol-, polyol-, zwitterionic sulfobetaine, or poly(2-sulphoethyl aspartamide) and other polar stationary phases chemically bonded on silica gel support but also in ion exchangers or zwitterionic materials showing combined HILIC–ion interaction retention mechanism. Some stationary phases are designed to enhance the mixed-mode retention character. Many polar columns show some contributions of reversed-phase (hydrophobic) separation mechanism, depending on the composition of the mobile phase, which can be tuned to suit specific separation problems [57].

Quantitative evaluation of the electrostatic interaction based on a systematic study of the nature and concentration of the salts in the mobile phase can be achieved for zwitterionic HILIC. A recent study showed that the separation of various nucleosides and nucleobases is based on a mechanism of partition and interaction through weak electrostatic forces and the contribution of the electrostatic interaction to the retention of the charged analytes may reach values of 25%–52% at low values of salt concentration. However, the electrostatic contribution decreased progressively as the salt concentration increases [61].

3.8 ELECTROSTATIC INTERACTIONS IN REVERSED-PHASE RETENTION MECHANISM

The reversed-phase high-performance liquid chromatography (RP-HPLC) is characterized by a mobile phase more polar than the stationary phase, and the main type of interactions in this HPLC technique are the hydrophobic interactions. However, some polar interactions (of electrostatic nature) are still possible, since many analytes having hydrophobic moieties also have ionic groups, and the stationary phase, typically silica, although derivatized with hydrophobic groups, also may show underivatized silanols that can act as weak acids [62,63]. Their interference in the retention mechanism is observable mainly for basic compounds, when disturbance on the peak shape are usually encountered [64,65]. Microcalorimetry is an instrumental technique that has been used for characterization of silanol activity by measuring the heat of adsorption of solvent molecules when they are immersed on stationary phase material. Different studies suggested that the measurement of solvent adsorption heat is governed by dipole–dipole interactions with residual silanols, which can be a useful tool for the determination of silanol activity and surface topography. For example, the heat of adsorption (ΔQ_{ads}) on the bare silica gel (with a concentration of silanol of 7.1 $\mu mol/m^2$) is proportional to the polarity of solvent, and the highest value was measured for methanol (around 56 J/g adsorbent), whereas for acetonitrile and hexane, ΔQ_{ads} was 33.8 J/g and 19.5 J/g, respectively. These results also show the contributions of different forces that are taking place between silanol and immersed solvent. Thus, methanol molecule can interact with silanol group by H-bond, which is stronger than dipole–dipole interactions between acetonitrile molecule and Si-OH. In the case of hexane, it can interact only due to the nonspecific dispersive forces. For bonded phases obtained from initial constant density of silanol groups (7.1 $\mu mol/m^2$), but with different coverage densities, a good correlation was observed between the heat of adsorption and coverage density with C18 chains. These correlations (based on data provided by Buszewski et al. [66]) are depicted in Figure 3.5, where a decrease in the heat of immersion with the increase in the coverage density can be observed. Also, on the low coverage density bonded phases, the heat of hexane immersion is lower than for methanol or acetonitrile. However, on the high-covered bonded phases, hexane adsorbs with higher heat (around 9 J/g) than acetonitrile (around 4.5 J/g), because this time the interactions with C18 chains are also important [65].

The interaction of the charges on a silica surface covered with the hydrophobic bonded phase with individual molecules in a medium may explain some exclusion processes observed in RP-LC. The typical phases in RP-HPLC consist of a silica surface covered with the hydrophobic bonded phase but also with a large number of silanol groups. The silanol groups have a slight acidic character and the ionization process leads to the accumulation on the silica surface of small negative charges. These charges may act through Coulombic forces toward ionic species such as naphthalene sulfonic acids (with structures and dipole moment simulation given in Figure 3.6) that are almost completely ionized in solution. This type of molecules is excluded from the pores of the stationary phase. For this reason, for example, the retention times for

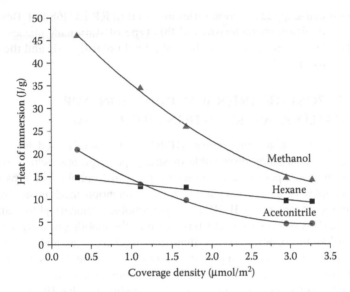

FIGURE 3.5   Correlations between heat of immersion and coverage density of silica surface with C18 chains for three solvents. ($y = ax^2 + bx + c$; MeOH: $a = 3.16 \times 10^{-2}$; $b = -1.966$; $c = 15.33$; $r^2 = 0.9563$; ACN: $a = 2.36$; $b = -19.6$; $c = 52.52$; $r^2 = 0.9978$; Hexane: 1.80; $b = -12.07$; $c = 24.66$; $r^2 = 0.9963$).

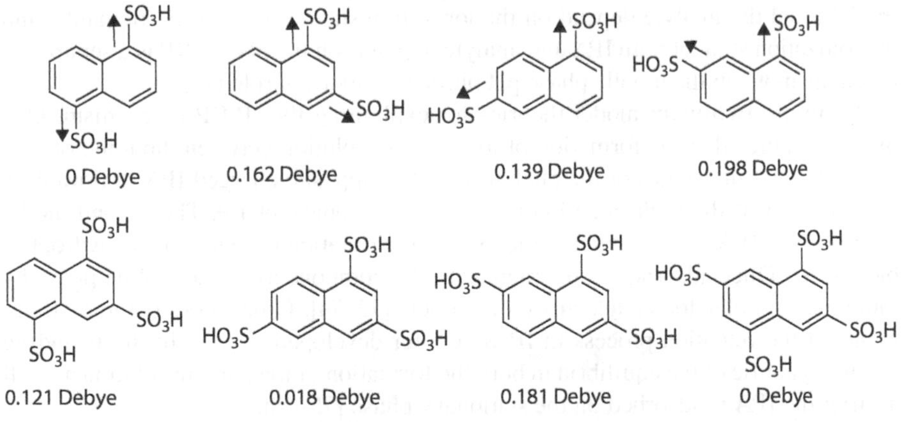

0 Debye   0.162 Debye   0.139 Debye   0.198 Debye

0.121 Debye   0.018 Debye   0.181 Debye   0 Debye

FIGURE 3.6   Simulation of dipole moments for eight naphthalene sulfonic acids used to characterize the silanol activity in RP-LC.

naphthalene sulfonic acids are shorter than holdup time (dead time) t_0 for the column, as measured with small compounds that are not retained but also are not excluded from the stationary phase pores (such as uracil or thiourea) [67].

Polar interactions are important and may interfere in the retention mechanism based on hydrophobic forces when stationary phases with embedded polar groups or with polar end-capped functionalities are used in RP-LC [68,69]. Besides that, the shape recognition characteristics of this type of stationary phases appear to be directly related to the density of the embedded polar ligands and the temperature of the separation [70].

3.9 ELECTROSTATIC INTERACTIONS IN ION-PAIR REVERSED-PHASE RETENTION MECHANISM

Ion pair in reversed-phase mechanism (IP-RP) is frequently used for the separation of analytes containing dissociable or strong polar groups, which make these compounds to have a poor interaction with typical hydrophobic columns (e.g., C18 or C8). Examples of analytes separated by this retention mechanism are organic acids, amino acids, or amines. IP-RP uses hydrophobic stationary phases and an ion-pairing agent (IPA or hetaeron), which is added to the mobile phase. IPA is selected such that it has the opposite charge than the target analytes and is able to form molecular association with. Compounds used as IPA can be, for example, a quaternary amine in case of separation of acids, or a strong organic acid (e.g., a sulfonic acid) in case of separation of amines. Another condition is that IPA must contain a hydrophobic moiety, which allows its bonding to the stationary phase through hydrophobic interactions. Overall, the effects of IPA on various analytes are as follows: (1) it enhances the retention of the analytes with the opposite charge to IPA and (2) it decreases the retention of analytes with the same charge with IPA and has a negligible effect on the retention of uncharged analytes. As the interactions between the IPA and the analyte depend on the ionization state of the two participants, and the ionization state of both IPA and analyte depends on the pH, IP-RP is a separation mechanism when the mobile phase pH plays an important role [71].

There are two main model theories for explaining the IP-RP mechanism. One model assumes that the formation of ion pairs in solution between the ionic or partially charged molecules of the analyte with the opposite charged IPA, followed by the retention of the preformed ion pair on the stationary phase. The second model assumes that IPA is first bound to the hydrophobic stationary phase by its hydrophobic moiety. Once adsorbed, it interacts with the components of the mobile phase by ionic and/or polar forces (electrostatic model) [72–74]. Other more complex treatments of the retention process in IP were later developed by Cecchi, by including thermodynamic of the equilibria in both the formation of ion pairs in solution as well as after the IPA is adsorbed on the stationary phase [75–78].

Electrostatic model assumes the formation of two parallel layers of charges (double electrostatic layer) when electrically charged molecules of IPA are adsorbed on the stationary phase surface. The adsorption of IPA is assumed to take place by hydrophobic interactions, and then the ionic groups of IPA will create the first adsorbed layer of charges. In this theory [79,80], only the first layer of charges is compact, whereas the second layer is diffuse and the electrical potential created by the first layer decreases exponentially away from the adsorbing surface (based on Gouy–Chapman theory of electrical double layer). The secondary diffuse layer is due to the IPA counterions (from mobile phase or provided by the solutes). Both layers are assumed to be under dynamic equilibrium. The counterions of opposite signs to those of the adsorbed IPA are electrostatically attracted toward the charged surface, and those with the same electrical charges are repelled from the surface [81].

Considering that IPA is a hydrophobic anion, IPA$^-$, the analyte is represented by the cation R-X$^+$, and the stationary phase is considered a ligand L, the main equilibria involved in this mechanism can be represented as follows:

$$L + IPA^- \rightleftarrows L * IPA^- \tag{3.37}$$

$$L * IPA^- + A\text{-}X^+ \rightleftarrows L * IPA^- \cdot {}^+X\text{-}R \tag{3.38}$$

According to the electrostatic model, the change in free standard enthalpy ΔG^0 involved into process can be written as having two contributions: one due to the hydrophobic interaction between IPA$^-$ and the hydrocarbon ligand L from stationary phase ($\Delta G^0_{hydrophobic}$) and the other due to the electrostatic interaction of analyte R-X$^+$ with charged surface of stationary phase resulted from the process (Equation 3.38) ($\Delta G^0_{electro}$):

$$\Delta G^0 = \Delta G^0_{hydrophobic} + \Delta G^0_{electro} \tag{3.39}$$

Taking into account the relation between capacity factor k' and equilibrium constant K_{IP} for the partition of the analyte between mobile and stationary phase, ($k'_{IP} = \Psi K_{IP}$, where Ψ is the phase ratio of the chromatographic column), the expression for k' can be written as follows:

$$k'_{IP} = \Psi \cdot \exp\left[-\frac{\Delta G^0_{hydrophobic} + \Delta G^0_{electro}}{RT}\right] = k'(0) \cdot \exp\left[-\frac{\Delta G^0_{electro}}{RT}\right] \tag{3.40}$$

where $k'(0)$ represents the retention factor of the analyte in the absence of IPA. The contribution to the retention factor due to the electrostatic interaction between charged analyte and the electric field created by IPA adsorbed onto the stationary phase is represented in this equation by the term $\exp(-\Delta G^0_{electro}/RT)$. This contribution represents the work involved in the transfer of an ionic analyte, generally having a charge z_i to the charged surface of the stationary phase is given by the formula:

$$\Delta G^0_{electro} = z_i F \Delta E \tag{3.41}$$

where:
   F represents the Faraday constant
   ΔE is the difference in electrostatic potential between the bulk of the mobile phase
      and the charged stationary phase surface

Thus, Equation 3.41 can be written as follows:

$$k'_{IP} = k'(0) \cdot \exp\left(\frac{z_i F \Delta E}{RT}\right) \tag{3.42}$$

The electrostatic potential ΔE resulted from the adsorption of IPA on stationary phase surface and its value can be obtained from Gouy–Chapman's theory. An expression for ΔE involving small charges and small electrical fields is the following [82,83]:

$$\Delta E = \frac{z_{IPA} n_{IPA}^{s.p.} F}{\kappa \varepsilon_0 \varepsilon_{m.p.}} \tag{3.43}$$

where:

$n_{IPA}^{s.p.}$ represents the surface molar fraction of the charged species (IPA)
ε_0 is the electrical permittivity of vacuum
$\varepsilon_{m.p.}$ is the dielectric constant of the mobile phase
κ is the Debye length given by the formula:

$$\kappa = F \left(\frac{\sum_j z_j^2 C_j}{\varepsilon_0 \varepsilon_{m.p.} RT} \right)^{1/2} \tag{3.44}$$

In this relation, the sum is made over all the ionic species found in the mobile phase. The electrostatic contribution to the change in free energy depends basically on the electric charge of the analyte and IPA involved into the retention process. Thus, the sign for the $\Delta G_{electro}^0$ can be positive, when the analyte and IPA have the same sign charges, and negative when they are of opposite charges.

Finally, the equation describing the dependence in IP electrostatic model of the capacity factor k_{IP}' on the main experimental parameters [80] has the general form:

$$\ln k_{IP}' = \ln k'(0) - \frac{z_i z_{IPA}}{z_{IPA}^2 + 1} \left(\ln C_{IPA} + \ln \frac{n_{IPA,max}^{s.p.} K_{IPA}}{\kappa} + \ln \frac{F^2}{\varepsilon_0 \varepsilon RT} + 1 \right) \tag{3.45}$$

where:

K_{IPA} represents the equilibrium constant for the adsorption of IPA on stationary phase
C_{IPA} is the molar concentration of IPA added to the mobile phase

According to this equation, the retention factor in IP-RP depends on several factors, such as follows:

- The retention factor of the analyte in the absence of IPA, $k'(0)$, which at its turn, depends on the hydrophobicity of analyte and the column phase ratio Ψ. However, the contribution of $k'(0)$ to the total value of k_{IP}' is supposed to be less significant compared to the other contributions, unless other interactions in the retention process are significant (e.g., interactions with residual silanols).
- There should be a linear dependence between $\ln k_{IP}'$ and $\ln C_{IPA}$, and the effect of charges of ionic species and IPA can be found in the slope with the value $(z_i z_{IPA}/(z_{IPA}^2 + 1))$. For example, if the ionic analyte and IPA have charges $+1$ and -1, respectively, the slope of the dependence of $\ln k_{IP}'$ on $\ln C_{IPA}$ is 0.5. Experimentally, it can be proven that the values of slope are around 0.5 [84,85].

- The effect of the nature and concentration of the organic modifier in mobile phase is found implicitly in the value of $k'(0)$ and K_{IPA}.
- The influence of the electrolyte concentration in the mobile phase is included in the expression of inverse Debye length κ, and the value of the capacity factor decreases with the increase of the electrolyte concentration in the mobile phase [86–88].
- The influence of pH on the value of retention factor is found in the term $\ln k'(0)$, and when the analyte becomes more dissociated by modifying the pH, its retention increases.

The electrostatic model is useful in explaining influence of many elution parameters on the retention of very polar or ionic species under IP-LC. However, the effect of IPA is only that of enhancing the retention, and less on selectivity, the interactions between the analyte and the stationary phase can be considered possible due to the addition of IPA.

3.10   ELECTROSTATIC INTERACTIONS IN ION CHROMATOGRAPHY

The mechanism of retention and separation in ion chromatography consists of ionic equilibria, and it is based on the difference in the affinity for the column of the ionic species that are separated. For the retention, the ions of the analyte should have higher affinity for the column than the counterions preexistent in the resin. The elution uses a mobile phase that contains competing ions (driving ions) that replace the analyte ions and *push* them out from the stationary phase. For an ion M^+ in the mobile phase, and a cation exchange in the form H^+, for example, the following equilibrium takes place:

$$\text{Resin-H} + M^+ \leftrightarrow \text{Resin-M} + H^+ \qquad (3.46)$$

This equilibrium is governed by a thermodynamic equilibrium constant K_{MH} given by the expression:

$$K_{MH} = [\text{Resin-M}]/[M^+]/[H^+]/[\text{Resin-H}] \qquad (3.47)$$

where [Resin-M] represents in fact the activity of M^+ in the resin, and [M^+] represents the activity of [M^+] in the mobile phase (similar notations for H^+), but the activities can be replaced with molar concentrations. Considering now the partition of species M^+ between the resin and the mobile phase, this is characterized by the distribution constant $K_d(M^+)$ given by the expression:

$$K_d(M^+) = K_{MH}[\text{Resin-H}]/[H^+] \qquad (3.48)$$

and the value for capacity factor k' for the separation becomes:

$$k'(M^+) = K_{MH}[\text{Resin-H}]/[H^+]\Psi \qquad (3.49)$$

where Ψ is the phase ratio for the chromatographic column.

The main interactions taking place in ion exchange are ionic interactions (although other types may be present). The ions that are exchanged are assumed distributed freely over the two phases: resin and solution. A boundary between the two phases can be visualized as a semipermeable membrane for all free ionic species (not for the acidic or basic functional groups that are ionic but are covalently connected to the skeleton of the resin). The ion exchange equilibrium can be in this case evaluated based on the Donnan membrane equilibrium theory. This theory refers to the uneven distribution of ions on the two sides of a semipermeable membrane separating solutions. When an electrolyte large M^+ is dissolved on one side of the membrane, and one ion is small enough to pass through the membrane while the other is too large or immobilized to penetrate the membrane, a difference in the electrical potential between the two sides of the membrane is generated. The expression for this potential is given by the formula (see [16]):

$$E_{\text{Donnan}} = \frac{1}{zF}\left(RT \ln \frac{a_{\text{resin}}}{a} - \Pi V\right) \qquad (3.50)$$

where:
   a_{resin} is the activity of the electrolyte in the resin
   a is the activity in the solution around the resin
   Π is the swelling (osmotic) pressure for the resin
   z is the charge of exchanging ions
   V is the partial molar volume of the ions
   F is the Faraday's constant

In an ion exchange process, there are two types of ions that are exchanged by the resin: H^+ and M^+, each creating a Donnan potential. At equilibrium, the two potentials must be equal. As a result, including all the necessary substitutions, the following expression for the equilibrium constant is obtained:

$$\ln K_{\text{MH}} = \frac{\Pi \cdot (V_{\text{H}} - V_{\text{M}})}{RT} \qquad (3.51)$$

A higher value of K_{MH} is obtained for smaller V_{M} and larger Π. When M has a charge higher than 1, K_{MH} is higher. The initial ions on the stationary phase (e.g., H^+) are typically the same as those used in the mobile phase as driving ions and the column is conditioned with the mobile phase before starting the separation. By increasing the concentration of these ions in the mobile phase, the elution is accelerated.

3.11   COMBINATIONS OF ELECTROSTATIC INTERACTIONS FOR LIQUID CHROMATOGRAPHY SEPARATIONS

Combination of electrostatic interactions takes place in case of chiral separations, when the chiral selector can be found on the surface of the stationary phase, or added to the mobile phase [89]. The separation is possible due to different interaction between the two enantiomers and the chiral selector, although short-ranged differences in intermolecular interactions of like and unlike pairs of chiral molecules

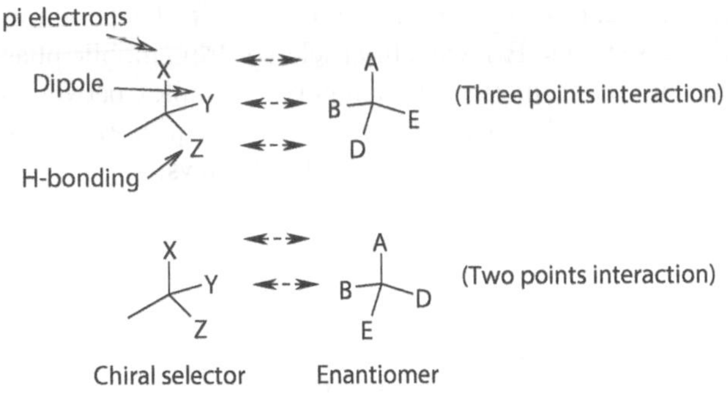

FIGURE 3.7   A simple model for illustrating the three points interaction principle for separating two enantiomers.

are typically smaller than the thermal energy [90]. The selector is able to discriminate between the two enantiomers if there are at least three points of electrostatic interactions between the chiral selector and one of the enantiomers as illustrated by Figure 3.7. Three points interaction principle (model), also designated as geometric model, is based on a simple observation that one of the enantiomers presents three groups that match exactly the three sites of the chiral selector from stationary phase, whereas its mirror image enantiomer can interact with the same selector with only two sites. In this way, the first enantiomer has a stronger interaction with chiral selector and elutes later than the second enantiomer. This simple model has not been developed as a quantitative model, and it does not predict which is the enantiomer that interacts with three points and which interacts with two points [91].

In case of asymmetrical cumulenes with three contiguous carbons linked together through double bonds, this principle is no longer applied. The enantiomers of asymmetrical cumulenes can interact differently in two points with cellulose carbamate-based stationary phase, and finally they are separated. Spectroscopic studies using infrared (IR) and vibrational circular dichroism (VCD), combined with density functional theory (DFT) calculations revealed that, in the presence of heptane, the stationary phase undergoes a conformational change due to intermolecular H-bonding between the $C=O$ and $N-H$ of the neighboring polymer chains [92].

However, empirical relationship between the chromatographic response (retention factor, k', or enantioselectivity, α_{DL}) with mobile phase composition in both reversed-phase and normal-phase chiral stationary phase liquid chromatography was noticed. Similarly to HILIC mechanism, the empirical relationships are described by the equations:

$$\ln k' = a + b \ln C_o + c C_o \tag{3.52}$$

$$\ln \alpha_{DL} = a' + b' \ln C_o \tag{3.53}$$

However, the effect of mobile phase composition on the selectivity of enantiomers D and L in RP-LC using chiral stationary phases is almost negligible [93].

When the chiral selector is added in mobile phase, this can interact differently with the two enantiomers in the mobile phase leading to their separation. A general model describing the competitive equilibria as a result of interaction (association) between enantiomers (E_1 and E_2) with chiral selector from mobile phase (denoted by P) takes into account the formation of a temporary complex between them ($E \times P$), which participates to the LC partition process. These equilibria and the correspondent thermodynamic constants are summarized as follows:

$$E_{1,m} \rightleftarrows E_{1,s} \quad K_{E1} \tag{3.54}$$

$$E_{2,m} \rightleftarrows E_{2,s} \quad K_{E2} \tag{3.55}$$

$$E_{1,m} + P_m \rightleftarrows (E_1 \times P)_m \quad K_{1,assoc} \tag{3.56}$$

$$E_{2,m} + P_m \rightleftarrows (E_2 \times P)_m \quad K_{2,assoc} \tag{3.57}$$

$$(E_1 - P)_m \rightleftarrows (E_1 \times P)_s \quad K_{d,1} \tag{3.58}$$

$$(E_2 - P)_m \rightleftarrows (E_2 \times P)_s \quad K_{d,2} \tag{3.59}$$

As the two enantiomers cannot be separated on an achiral stationary phase, $K_{E1} = K_{E2}$. Usually, $K_{d,1}$ is also identical with $K_{d,2}$, and thus the chiral separation cannot be obtained from different partitions of the two *complexes* between mobile and stationary phases. Nevertheless, this can be achieved only from the difference of the stability constants of the *complexes* formed between enantiomers and the chiral selector from mobile phase. If $K_{1,assoc} > K_{2,assoc}$ then the order of elution is $t_{R,1} > t_{R,2}$, and enantioselectivity is given by the ratio between the values of $K_{1,assoc}$ and $K_{2,assoc}$.

Examples of inclusion complexes that can be formed in mobile phase are those based on the addition of free cyclodextrin into the aqueous component [94–96]. Due to the hydrophilic outer surface, they will not interact with hydrophobic stationary phase significantly, but its hydrophobic inner surface is able to interact with hydrophobic moiety of analytes. This interaction can be different for the two enantiomers, finally leading to their separation. According to the aforementioned model, the order of elution is now reversely compared to chiral separation when cyclodextrin is bound to a stationary phase support (which is equivalent to a C4 or C8 stationary phase). The mobile phase contains an aqueous component and organic additives, such as acetonitrile, methanol, ethanol, or 2-propanol. Ethanol and 2-propanol exhibit greater affinity for the cyclodextrin cavity and will displace solutes to a greater degree than methanol or acetonitrile [97,98]. One major problem is, however, the elimination of residual cyclodextrin from the column before using it for another purposes rather than chiral separations.

Another example is the use of sulfated β-cyclodextrin (*S*-β-CD) as an additive in the mobile phase for the enantiomeric separation of some chiral aromatic amines [99]. In this case the process is more complicated and a combination of intermolecular forces can be used to describe the separation process. VCD experiments indicate that the interactions of the two enantiomers with the *S*-β-CD occur through an inclusion of the aromatic part of the analyte, as well as through electrostatic interaction between the protonated amine and the sulfate groups located at the narrow part of the *S*-β-CD. This study showed that the N atom of each enantiomer is structurally close

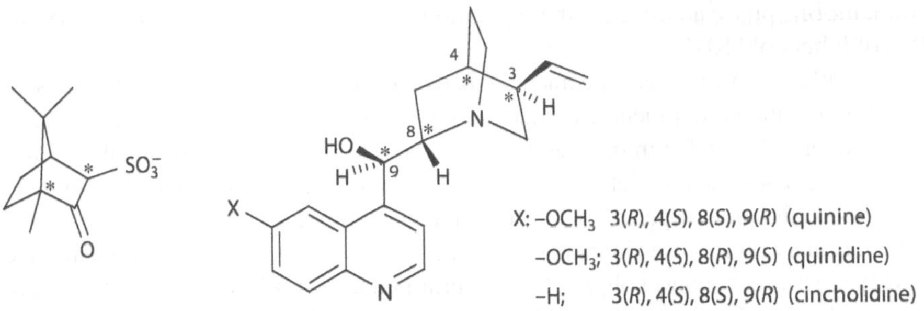

X: –OCH₃ 3(R), 4(S), 8(S), 9(R) (quinine)

–OCH₃; 3(R), 4(S), 8(R), 9(S) (quinidine)

–H;   3(R), 4(S), 8(S), 9(R) (cincholidine)

FIGURE 3.8 Structure of some quinine-related compounds used in chiral separation based on ion-pairing mechanism.

to the sulfate groups, at a distance of 3.58 Å and 3.63 Å for the R- and S-enantiomers, respectively. In addition, the phenyl groups of both enantiomers are included in the cavity of S-β-CD. The R-enantiomer is slightly closer to the interior wall of the annulus than the S-enantiomer. These results suggest that the interaction between the R-enantiomer of studied amines and the S-β-CD is stronger than for the S-enantiomer.

Complexes between enantiomers and chiral selector in mobile phase can also be formed by electrostatic interactions, which is a process similar to the ion pairing. Separation of enantiomers by ion-pairing mechanism is possible using a chiral counterion found in mobile phase that binds electrostatically the enantiomeric molecules leading to diastereoisomeric ion pairs [100,101]. They can then be separated on polar or hydrophobic stationary phases depending on the nature of chiral counterion and the composition of the mobile phase. The binding forces are usually a combination of electrostatic attraction and H-bonding, and for this reason the mobile phase of low polarity is normally used to favor the formation of ion pair. The selection of the chiral counterion is often made according to the *three points rule* (as schematically described in Figure 3.8), which states that an interaction in three points between chiral agent and at least one of the enantiomers is necessary to obtain stereoselective retention. For example, the two enantiomers of propranolol can be separated using mobile phase containing tartaric acid as counterion in a buffered aqueous mobile phase [102]. Dibutyl ester of tartaric acid is also known for separation of enantiomers of propranolol and other aminoalcohols on phenyl silica stationary phase using buffered aqueous mobile phases containing hexafluorophophate. Enantiomers of amino alcohols can also be separated on diol stationary phases using quinine congeners (in mobile phase containing dichloromethane/1-pentanol), or (+)-10-camphorsulphonic acid (in mobile phase consisting in dichloromethane/1-pentanol). In these cases, the addition of a weak acid (formic or acetic) in order to protonate the analyte enantiomers or the chiral counterion is also recommended [103].

The addition of albumin in buffered aqueous mobile phase can also be used for separation of enantiomers of carboxy containing compounds in their dissociated form, under conditions of RP mechanism [100]. Another example of complex chiral separation is the use of double-helical chelate, namely (–)(M)(λ,Λ)-[4,4′-(1S)-methyl-(2R)-propylethane-diyldi-imino) bis(pent-3-en-2-onato)]nickel(II) as the chiral mobile phase additive for the separation of four stereoisomers (*SR*, *RS*, *SS*, and *RR*) of labetalol [104].

A combination of several interactions between analyte and stationary phase is the base for so-called biomimetic chromatography [105,106]. Biomimetic HPLC is based on stationary phases that mimic the lipid environment of a membrane, and they can be used in the assessment of interactions between molecules and biological membranes or for the estimation of permeability through cell membranes. Examples of new available stationary phases used in biomimetic HPLC are immobilized artificial membranes (IAM) consisting of phospholipids, human serum albumin (HSA), and α-acid glycoprotein (AGP) chemically bonded on silica skeleton [107–109]. In both cases, the driving forces responsible for the interaction between analyte and stationary phase are electrostatic forces (ion–dipole, dipole–dipole, H-bonding), although van der Waals forces can interfere in the separation mechanism due to the hydrophobic part (linker) from stationary phase and hydrophobic groups from analyte molecule [110,111]. An example of stationary phase used in biomimetic liquid chromatography is illustrated in Figure 3.9, which contains functionalities with hydrophobic and hydrophilic character [108].

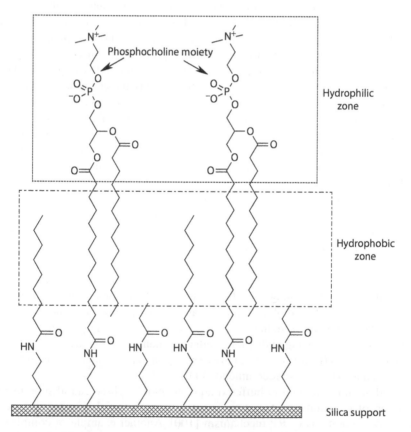

FIGURE 3.9 Chemistry of surface for biomimetic stationary phase based on hydrophilic/hydrophobic distinct zones.

3.12   SOLVOPHOBIC THEORY

The solvophobic theory of Sinanoglu [112] was applied by Horvath to the retention process in RP-LC to describe this process as caused by the energy gain when the weak interactions of a hydrophobic solute with the polar solvent are replaced with interactions only between the polar solvent molecules [113]. According to this model, the stationary phase acts as a passive receptor to hydrophobic molecules that are repelled by the aqueous mobile phase. The mobile phase is treated as a bulk, and the interaction between analyte and mobile phase has the Onsager model as background, applied to the solvation processes. Also, specific interactions between the residual silanols from stationary phase and the polar groups from solute molecule are not taken into consideration.

The equilibrium constant of a studied compound A in a partition process (K_i) between two immiscible phases (denoted by α and β) is dependent on the variation of free standard enthalpy (ΔG_i^0) according to the Nernst equation:

$$K_A = e^{\frac{\Delta G_A^0}{RT}} \tag{3.60}$$

Assuming that no volume changes occur during partition of a molecule from one phase to another, the free enthalpy ΔG_i^0 can be replaced by the free energy ΔE_A^0, which is the difference of energies at the transfer of A from phase α to phase β, namely ($E_{A,\beta}^0 - E_{A,\alpha}^0$). The free energies $E_{A,\alpha}^0$ and $E_{A,\beta}^0$ represent the energy necessary to place the molecule A from vacuum to a solution formed by molecules of the phase α and β, as a simplified model depicted in Figure 3.10. Basically, this energy is required to create a cavity in the bulk of solvent α or β (E_A^{cav}), where the molecule A interacts with the solvating molecules by van der Waals forces with

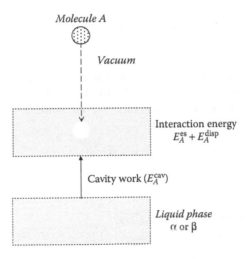

FIGURE 3.10   Model for calculation of free energy at the transfer of solute molecule from vacuum into the solvent bulk.

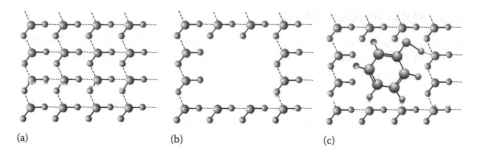

FIGURE 3.11 Model for the creation of a cavity in water by breaking hydrogen bonds and placing a phenol molecule in the cavity. (a) Water with H-bonds, (b) creation of the cavity, and (c) placing the phenol molecule.

energy denoted by E_A^{vdW}. A model for the creation of a cavity in water by breaking H-bonds and placing a phenol molecule as solute in the cavity is illustrated in Figure 3.11.

In this simplified model, the sum of the two energies gives the value of E_A^0, that is

$$E_A^0 = E_A^{\text{cav}} + E_A^{\text{vdW}} \tag{3.61}$$

(For simplicity, the index 0 for standard values and the index for the nature of solvating medium, α or β are omitted for different contributions to the total energy). Taking into consideration that the van der Waals energy E_A^{vdW} has two main contributions, caused by electrostatic forces E_A^{es} and by the dispersion forces E_A^{disp}, the expression of E_A^0 can be written as the sum:

$$E_A^0 = E_A^{\text{cav}} + E_A^{\text{es}} + E_A^{\text{disp}} \tag{3.62}$$

The expressions for these contributions to the total energy have been reported in literature [111–114] and applied to the partition process [115], and they are briefly discussed as follows. For example, the free energy for the cavity formation in the solvent where to accommodate the molecule A can be expressed by the formula [116]:

$$E_A^{\text{cav}} = \kappa_A^e S_A \gamma_i (1 - w_i) N \tag{3.63}$$

where:
S_A is the surface area of molecule A
γ is the surface tension of the liquid phase
N is Avogadro number
w_i is a correction factor
κ_A^e is a special energetic parameter depending on phase $i = \alpha, \beta$, and A, which are
 further explicated

The expression of κ_A^e is as follows:

$$\kappa_A^e = 1 + (\kappa_i^e - 1) \cdot \left(\frac{V_i}{V_A}\right)^{2/3} \tag{3.64}$$

where V_S and V_A are the molar volumes for the phase solvent i and for the species A, respectively, which can be calculated from the molecular weight and density ρ of the compound (i.e., $V_i = M_i/\rho_i$, and $V_A = M_A/\rho_A$); the energetic parameter κ_i^e corresponds to the pure solvent, which is given by the expression:

$$\kappa_i^e = \frac{N^{1/3} \Delta H_i^{\text{vap}}}{V_i^{2/3} \gamma_i \left(1 - \dfrac{d \ln \gamma}{d \ln T} - \dfrac{2C_{\exp,i}T}{3}\right)} \tag{3.65}$$

where:
 T is the absolute temperature
 ΔH_i^{vap} represents the evaporation heat for the phase solvent i, whose values for common solvents can be found in literature [117]

The parameter w_i is given by the following dependence on energetic parameter κ_i^e and on the entropic one, denoted by κ_i^s and having a similar relationship described in Equation 3.64:

$$w_i = \left(1 - \frac{\kappa_A^s}{\kappa_A^e}\right)\left(\frac{d \ln \gamma_i}{d \ln T} + \frac{2}{3} C_{\exp,A}T\right) \tag{3.66}$$

With aid of these relationships, the expression for the E_A^{cav} free energy component becomes:

$$E_A^{\text{cav}} = \left[1 - (\kappa_S^e - 1)\left(\frac{V_S}{V_A}\right)^{2/3}\right] \cdot N \cdot A \cdot \gamma \cdot (1 - w_S) \tag{3.67}$$

The expression for E_A^{es} is obtained from the Onsager reaction field [37] and is given by the formula:

$$E_A^{\text{es}} = -\frac{N^2 \mu_A^2}{2V_A} D_S P_{A,S} \tag{3.68}$$

where the parameters D_S and $P_{A,S}$ are given by

$$D_S = \frac{2(\varepsilon_S - 1)}{2\varepsilon_S + 1} \tag{3.69}$$

$$P_{A,S} = \frac{V_A}{4\pi\varepsilon_0(V_A - N\, D_S \alpha_A)} \tag{3.70}$$

where:

μ_A is the dipole moment of solute molecule

α_A is the polarizability of analyte

ε_s is the dielectric constant of solvent S

The expression of the dispersion term E_A^{dip} can be obtained using an effective pair potential that contains a correction from the gas phase potential of the interaction of Lennard-Jones, and a final expression for this energetic contribution is written as follows [115]:

$$E_A^{dis} = -\frac{15.23}{8\pi}(\Omega'_{A,S} + \Omega''_{A,S})\gamma D_A D_S \qquad (3.71)$$

In this dependence, the parameters $\Omega'_{A,S}$ and $\Omega''_{A,S}$ depend on the diameter of solute molecule and solvent molecule (and in a good approximation it is taken $\Omega''_{A,S} = 0.1 \cdot \Omega'_{A,S}$), on Kihara parameter [118], the parameters D_A and D_S are calculated from Clausius-Mosotti function.

Applied to the partition process of a solute A from phase α to phase β, a general expression for K_A based on solvophobic theory is found in Equation 3.72 [16]:

$$\log K = a\,A_A + b\,V_A^{2/3}A_A + c\,\mu_A^2 + d\,\alpha_A \qquad (3.72)$$

where:

A_A represents the van der Waals molecular surface area of molecule A

V_A is the molar volume

μ_A is the dipole moment

α_A is the polarizability

The values for the parameter a, b, c, and d depend on a variety of molecular parameters such as molar volumes, molecular diameters, critical pressures, Kihara parameters, and ionization potentials for molecule A and solvent S, and on surface tension and dielectric constant of solvent, which can be found in literature. A possible application of the mathematical framework of this theory to the calculation of the phase ratio, Ψ, was already reported, and a good agreement with other approaches was observed [119,120]. As indicated by Equation 3.65, a linear correlation should be observed between $\log K$ and van der Waals molecular surface area. An example is given in Figure 3.11, where a good correlation can be noticed for the dependence between the 10-base logarithm of the octanol/water partition constant ($\log K_{ow}$) and van der Waals molecular surface area, calculated with addition of MarvinSketch program (Figure 3.12) [12].

Another example of application of solvophobic theory is the explanation of lower solubility in water of perfluorinated alkanes compared to the corresponding hydrocarbons. The polarity of C–F bonds is much higher than that of C–H bonds, and consequently a stronger interaction with the water molecules is expected to take place. According to the energetic model previously presented, the energy for cavity formation has a significant contribution. Fluorocarbons have a larger molar volume (and molecular surface area) than the corresponding hydrocarbons. For example, to a

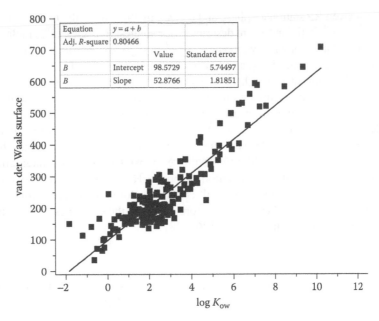

FIGURE 3.12   Linear correlation between theoretical log K_{ow} and van der Waals area for 200 organic compounds computed by MarvinSketch program.

fluorocarbon with a surface area of 299.9 Å^2 (perfluorohexane) corresponds a hydrocarbon (hexane) with surface area of 215.6 Å^2. The work done to form a cavity large enough to accommodate a fluorocarbon (E^{cav}) offsets the anticipated free-energy benefit from enhanced energetic interactions with water. An entropic effect due to movement restrictions imposed to the solvent molecules by the polar fluorocarbon may also contribute to the low fluorocarbon solubility in water [121].

3.13 CONCLUSIONS

The evaluation of molecular interactions is important for the understanding, and in some cases, in predicting the directional behavior of solutes in a specific separation, the quantitative results regarding the separation are not always accurate. This is caused mainly by the complexity of the chromatographic process and by the approximations frequently used to evaluate molecular interactions and properties. Instead, alternative paths are more often employed for obtaining empirical dependences between chromatographic property and values of important physico-chemical or molecular parameters of the participants in the separation process (solutes, mobile phase, and stationary phase) [122]. A common approach is based on linear free energy relationships [123], which are used for the prediction of a given chromatographic property P_r as a dependence on several physico-chemical or molecular characteristics, such as in the following model applied in reversed-phase mechanism:

$$\log P_r = c + r\, R_2 + s\, \pi_2^H + a \sum \alpha_2^H + b \sum \beta_2 + v\, V_x \tag{3.73}$$

where for a given RP stationary phase and a given mobile phase, the molecular parameters R_2, π_2^H, $\sum \alpha_2^H$, $\sum \beta_2$, and V_x are descriptors for solutes: R_2 is the excess molar refraction, π_2^H is the solute polarizability/dipolarity, $\sum \alpha_2^H$ and $\sum \beta_2$ are the solute hydrogen-bond acidity and basicity, respectively, and V_x is the molecular volume of solute [124,125].

Another direction of research in molecular interactions of the chromatographic process is its molecular simulation, which is based on molecular dynamics and Monte Carlo methods that have been developed and applied mainly to the modeling of RP-HPLC. These approaches are focused on molecular-level understanding of the retention mechanism by studying the structure and dynamics of the bonded phase and its interface with the mobile phase, as well as the main interactions of solutes with the active sites from stationary phase. However, these investigations of the retention process pose many difficult challenges for computations, for example, the accuracy of the molecular mechanics force fields or the efficiency of the sampling algorithms used in these simulations [126–128].

REFERENCES

1. D. J. Bornhop, M. N. Kammer, A. Kussrow, R. A. Flowers, and J. Meiler, Origin and prediction of free-solution interaction studies performed label-free, *PNAS, 113*: E1595–E1604, 2016.
2. T. Hanai, *In silico* modeling study on molecular interactions in reversed-phase liquid chromatography, *J. Chrom. Sci., 53*: 1084–1091, 2015.
3. E. Forgacs and T. Cserhati, Chapter 1: Adsorption phenomena and molecular interactions in chromatography; in *Molecular Basis of Chromatographic Separation*, CRC Press, Boca Raton, FL, pp. 3–14, 1997.
4. J. A. Yancey, Review of liquid phases in gas chromatography; Part I: Intermolecular forces, *J. Chromatogr. Sci., 32*: 349–357, 1994.
5. O. Sinanoglu, Intermolecular forces in liquids; in *Advances in Chemical Physics* (J. O. Hirschfelder, Ed.), Vol. 12, John Wiley & Sons, New York, p. 283, 1967.
6. A. Salam, *Molecular Quantum Electrodynamics: Long-Range Intermolecular Interactions*, Wiley, New York, 2010.
7. A. Stone, *Theory of Intermolecular Forces*, 2nd edition, Oxford University Press, Oxford, 2013.
8. G. V. Gibbs, T. D. Crawford, A. F. Wallace, D. F. Cox, R. M. Parrish, E. G. Hohenstein, and C. D. Sherrill, Role of long-range intermolecular forces in the formation of inorganic nanoparticle clusters, *J. Phys. Chem. A, 115*: 12933–12940, 2011.
9. P. Hobza and R. Zahradnik, *Weak Intermolecular Interactions in Chemistry and Biology*, Elsevier Scientific Publishing Company, Amsterdam, the Netherlands, p. 13, 1980.
10. E. E. Meyer, K. J. Rosenberg, J. Israelachvili, Recent progress in understanding hydrophobic interactions, *PNAS, 103* (43): 15739–15746, 2006.
11. E. Bacalum, T. Galaon, V. David, and H. Y. Aboul-Enein, Retention behavior of some compounds containing polar functional groups on perfluorophenylsilica based stationary phase, *Chromatographia, 77*: 543–552, 2014.
12. Marvin Beans Software: Marvin View, Marvin Space, Marvin Sketch with calculator plugins, version 5.2.5.1., ChemAxon Ltd.
13. E. Bacalum, M. Tanase, M. Cheregi, H. Y. Aboul-Enein, and V. David, Retention mechanism in zwitterionic hydrophilic interaction liquid chromatography (ZIC-HILIC) studied for highly polar compounds under different elution conditions, *Rev. Roum. Chim., 61*: 531–539, 2016.

14. V. David and S. Moldoveanu, Interaction between two charged distributions situated into a continuum, infinite and isotropic medium, *Rev. Roum. Chim.*, *26*: 333–340, 1981.
15. H. Margenau and N. R. Kestner, *Theory of Intermolecular Forces*, 2nd edition, Pergamon Press, Oxford, p. 16, 1971.
16. S. C. Moldoveanu and V. David, *Essentials in Modern HPLC Separations*, Elsevier, Amsterdam, the Netherlands, p. 115, 2013.
17. K. J. Miller and J. Savchik, A new empirical method to calculate average molecular polarizabilities, *J. Am. Chem. Soc.*, *101*: 7206–7213, 1979.
18. K. J. Miller, Calculation of the molecular polarizability tensor, *J. Am. Chem. Soc.*, *112*: 8543–8551, 1990.
19. S. Moldoveanu and A. Savin, Aplicatii in Chimie ale Metodelor Semiempirice de Orbitali Moleculari, Edit. Academiei RSR, Bucuresti, Romania, 1980.
20. E. Talebian and M. Talebian, A general review on the derivation of Clausius-Mossotti relation, *Optik*, *124*: 2324–2326, 2013.
21. M. E. Davis, J. D. Madura, B. A. Luty, and J. A. McCammon, Electrostatics and diffusion of molecules in solutions: Simulation with the University of Huston Brownian dynamics program, *Comp. Phys. Comm.*, *62*: 187–197, 1991.
22. M. E. Davis and J. A. McCammon, Calculating electrostatic forces from grid-calculated potentials, *J. Comput. Chem.*, *11*: 401–409, 1990.
23. O. Guvench and C. L. Brooks, Efficient approximate all-atom solvent accessible surface area method parameterized for folded and denatured protein conformations, *J. Comput. Chem.*, *25*: 1005–1014, 2004.
24. W. C. Still, A. Tempczyk, R. C. Hawley, and T. Hendrickson, Semianalytical treatment of solvation for molecular mechanics and dynamics, *J. Am. Chem. Soc.*, *112*: 6127–6129, 1990.
25. B. Buszewski, R. Gadzała-Kopciuch, and M. Michel, Thermodynamic description of retention mechanism in liquid chromatography, *Accred. Qual. Assur.*, *16*: 237–244, 2011.
26. J. Israelachvili, *Intermolecular and Surface Forces*, Academic Press, Amsterdam, the Netherlands, 1991.
27. Y. Marechal, *The Hydrogen Bond and the Water Molecule*, Elsevier, Amsterdam, the Netherlands, p. 6, 2007.
28. R. Kaliszan, *Quantitative Structure-Chromatographic Retention Relationship*, John Wiley & Sons, New York, 1987.
29. N. Tanaka, K. Kimata, K. Hosoya, H. Miyanishi, and T. Araki, Stationary phase effects in reversed-phase liquid chromatography, *J. Chromatogr. A*, *656*: 265–287, 1993.
30. H. Engelhardt and M. Jungheim, Comparison and characterization of reversed phases stationary phases, *Chromatographia*, *29*: 59–68, 1990.
31. E. Bacalum, M. Cheregi, and V. David, Retention behaviour of some benzodiazepines in solid-phase extraction using modified silica adsorbents having various hydrophobicities, *Rev. Roum. Chim.*, *60*: 891–898, 2015.
32. M. Mifune, Y. Shimomura, Y. Saito, Y. Mori, M. Onoda, A. Iwado, N. Motohashi, and J. Haginaka, High-performance liquid chromatography stationary phases based on π–π electron interaction. Aminopropyl silica gels modified with metal phthalocyanines, *Bull. Chem. Soc. Jap.*, *71*: 1825–1829, 1998.
33. K. Croes, A. Steffens, D. H. Marchand, and L. R. Snyder, Relevance of pi-pi and dipole-dipole interactions for retention on cyano and phenyl columns in reversed-phase liquid chromatography, *J. Chromatogr. A*, *1098*: 123–130, 2005.
34. M. Yang, S. Fazio, D. Munch, and P. Drumm, Impact of methanol and acetonitrile on separations based on pi-pi interactions with a reversed-phase phenyl column, *J. Chromatogr. A*, *1097*: 124–129, 2005.

35. M. Mnatsakanyan, P. G. Stevenson, D. Shock, X. A. Conlan, T. A. Goodie, K. N. Spencer, N. W. Barnett, P. S. Francis, and R. A. Shalliker, The assessment of π-π selective stationary phases for two-dimensional HPLC analysis of foods: Application to the analysis of coffee, *Talanta*, *82*: 1349–1357, 2010.

36. T. Galaon, E. Bacalum, M. Cheregi, and V. David, Retention studies for large volume injection of aromatic solvents on phenyl-silica based stationary phase in RP-LC, *J. Chromatogr. Sci.*, *51*: 166–172, 2013.

37. L. Onsager, Electric moments of molecules in liquids, *J. Amer. Chem. Soc.*, *58*: 1486–1493, 1936.

38. J. G. Kirkwood, Statistical mechanics of liquid solutions, *Chem. Rev.*, *19*: 275–307, 1936.

39. J. G. Kirkwood, Order and disorder in liquid solutions, *J. Chem. Phys.*, *43*: 97–107, 1939.

40. J. G. Kirkwood, Theoretical studies upon dipolar ions, *Chem. Rev.*, *24*: 233–251, 1939.

41. D. L. Beveridge and G. W. Schnuelle, Free energy of a charge distribution in concentric dielectric continua, *J. Phys. Chem.*, *79*: 2562–2566, 1975.

42. D. L. Beveridge and G. W. Schnuelle, Statistical thermodynamic supermolecule-continuum study of ion hydration. I. Site method, *J. Phys. Chem.*, *79*: 2566–2573, 1975.

43. F. Deanda, K. M. Smith, J. Liu, and R. S. Pearlman, GSSI, a general model for solute-solvent interactions. 1. Description of the model, *Mol. Pharm.*, *1*: 23–39, 2004.

44. R. S. Pearlman, Molecular surface areas and volumes and their use in structure/activity relationships, in *Physical Properties of Drugs* (S. H. Yalkowsky, A. A. Sinkula and S. C. Valvani, eds.), Marcel Dekker, New York, pp. 321–347, 1980.

45. D. Eisenberg and A. D. McLachlan, Solvation energy in protein folding and biding, *Nature*, *319*: 199–203, 1986.

46. T. Lazaridis and M. Karplus, Effective energy function for proteins in solution, *Proteins*, *35*: 133–152, 1999.

47. P. Ferrara, J. Apostolakis, and A. Caflisch, Evaluation of a fast implicit solvent model for molecular dynamics simulations, *Proteins*, *46*: 24–33, 2002.

48. M. L. Connolly, Solvent-accessible surfaces of proteins and nucleic-acids, *Science*, *221*: 709–713, 1983.

49. E. Durham, B. Dorr, N. Woetzel, R. Staritzbichler, and J. Meiler, Solvent accessible surface area approximations for rapid and accurate protein structure prediction, *J. Mol. Model.*, *15*: 1093–1108, 2009.

50. S. A. Ali, M. I. Hassan, A. Islam, and F. Ahmad, A review of methods available to estimate solvent-accessible surface areas of soluble proteins in the folded and unfolded states, *Curr. Protein Pept. Sci.*, *15*: 456–476, 2014.

51. T. Cserhati, Multivariate Methods in Chromatography: A Practical Guide, John Wiley & Sons, Chichester, p. 6, 2008.

52. E. Deconinck, T. Verstraete, E. Van Gyseghem, Y. Vander Heyden, and D. Coomans, Orthogonal chromatographic descriptors for modelling Caco-2 drug permeability, *J. Chromatogr. Sci.*, *50*: 175–183, 2012.

53. J. Kujawski, H. Popielarska, A. Myka, B. Drabińska, and M. K. Bernard, The log P parameter as a molecular descriptor in the computer-aided drug design—An overview, *Comput. Meth. Sci. Technol.*, *18*: 81–88 (2012).

54. S. Bocian, Solvation processes in liquid chromatography: The importance and measurements, *J. Liq. Chromatogr. Relat. Technol.*, *39*: 731–738, 2016.

55. S. Bocian, A. Felinger, and B. Buszewski, Comparison of solvent adsorption on chemically bonded stationary phases in RP-LC, *Chromatographia*, *68*: S19–S26, 2008.

56. G. Greco and T. Letzel, Main interactions and influences of the chromatographic parameters in HILIC separations, *J. Chromatogr. Sci.*, *51*: 684–693, 2013.

57. P. Jandera, Stationary and mobile phases in hydrophilic interaction chromatography: A review, *Anal. Chim. Acta*, *692*: 1–25, 2011.
58. D. V. McCalley, Is hydrophilic interaction chromatography with silica columns a viable alternative to reversed-phase liquid chromatography for the analysis of ionisable compounds? *J. Chromatogr. A*, *1171*: 46–55, 2007.
59. E. Karatapanis, Y. C. Fiamegos, and C. D. Stalikas, A revisit to the retention mechanism of hydrophilic interaction liquid chromatography using model organic compounds, *J. Chromatogr. A*, *1218*: 2871–2879, 2011.
60. Y. Liu and A. Vailaya, Normal-phase HPLC, in HPLC for *Pharmaceutical Scientists*, (Y. Kazakevich and R. LoBrutto, eds.), John Wiley & Sons, Hoboken, NJ, p. 241, 2007.
61. D. García-Gómez, E. Rodríguez-Gonzalo, and R. Carabias-Martínez, Evaluation of the electrostatic contribution to the retention of modified nucleosides and nucleobases by zwitterionic hydrophilic interaction chromatography, *ISRN Anal. Chem.*, article 308062, 2012.
62. J. J. Kirkland, Development of some stationary phases for reversed-phase high-performance liquid chromatography, *J. Chromatogr. A*, *1060*: 9–21, 2004.
63. M. J. Wirth and H. O. Fatunmbi, Horizontal polymerization of mixed trifunctional silanes on silica. 2. Application to chromatographic silica gel, *Anal. Chem.*, *65*: 822–826, 1993.
64. D. V. McCalley, Evaluation of reversed-phase columns for the analysis of very basic compounds by high-performance liquid chromatography: Application to the determination of the tobacco alkaloids, *J. Chromatogr. A*, *636*: 213–220, 1993.
65. S. Bocian and B. Buszewski, Residual silanols at reversed-phase silica in HPLC—a contribution for a better understanding, *J. Sep. Sci.*, *35*: 1191–1200, 2012.
66. B. Buszewski, S. Bocian, and G. Rychlicki, Investigation of silanol activity on the modified silica surfaces using microcalorimetric measurements, *J. Sep. Sci.*, *34*: 773–779, 2011.
67. P. Jandera, S. Bunčeková, M. Halama, K. Novotná, and M. Nepraš, Naphthalene sulphonic acids—new test compounds for characterization of the columns for reversed-phase chromatography, *J. Chromatogr. A*, *1059*: 61–72, 2004.
68. G. P. O'Sullivan, N. M. Scully, and J. D. Glennon, Polar-embedded and polar-end capped stationary phases for LC, *Anal. Lett.*, *43*: 1609–1629, 2010.
69. J. Layne, Characterization and comparison of the chromatographic performance of conventional, polar-embedded, and polar-endcapped reversed-phase liquid chromatography stationary phases, *J. Chromatogr. A*, *957*: 149–164, 2002.
70. C. A. Rimmer and L. C. Sander, Shape selectivity in embedded polar group stationary phases for liquid chromatography, *Anal. Bioanal. Chem.*, *394*: 285–291, 2009.
71. T. Cecchi, *Ion-Pair Chromatography and Related Techniques*, CRC Press, Boca Raton, FL, 2010.
72. C. Horvath and S. R. Lipsky, Use of liquid ion exchange chromatography for the separation of organic compounds. *Nature*, *211*: 748–749, 1966.
73. C. Horvath, W. Melander, I. Molnar, and P. Molnar, Enhancement of retention by ion-pair formation in liquid chromatography with nonpolar stationary phases, *Anal. Chem.*, *49*: 2295–2305, 1977.
74. J. H. Knox and R. A. Hartwick, Mechanism in ion-pair liquid chromatography of amines, neutrals, zwitterions, and acids using anionic hetaerons, *J. Chromatogr. A*, *204*: 2–21, 1981.
75. T. Cecchi, F. Pucciarelli, and P. Passamonti, Extended thermodynamic approach to ion interaction chromatography, *Anal. Chem.*, *73*: 2632–2639, 2001.
76. T. Cecchi, Extended thermodynamic approach to ion interaction chromatography. Influence of the chain length of the solute ion: A chromatographic method for the determination of ion-pairing constants, *J. Sep. Sci.*, *29*: 549–554, 2005.

77. T. Cecchi, F. Pucciarelli, and P. Passamonti, Extended thermodynamic approach to ion interaction chromatography. A mono-and bivariate strategy to model the influence of ionic strength, *J. Sep. Sci.*, *27*: 1323–1332, 2004.

78. T. Cecchi, Use of lipophilic ion adsorbtion isotherms to determine the surface area and the monolayer capacity of a chromatographic packing as well as the thermodynamic equilibrium constant for its adsorption, *J. Chromatogr. A*, *1072*: 201–206, 2005.

79. J. Ståhlberg, A quantitative evaluation of the electrostatic theory for ion pair chromatography, *Chromatographia*, *24*: 820–826, 1987.

80. A. Bartha and J. Ståhlberg, Electrostatic model of the reversed-phase ion-pair chromatography, *J. Chromatogr. A*, *668*: 255–284, 1994.

81. T. Cecchi, Ion pairing chromatography, *Crit. Rev. Anal. Chem.*, *38*: 161–213, 2008.

82. T. Cecchi and P. Passamonti, Retention mechanism for ion-pair chromatography with chaotropic reagents, *J. Chromatogr. A*, *1216*: 1789–1797, 2009.

83. T. Cecchi, Theoretical models of ion pair chromatography: A close up of recent literature production, *J. Liq. Chromatogr. Rel. Technol.*, *38*: 404–414, 2015.

84. M. Radulescu and V. David, Partition versus electrostatic model applied to the ion-pairing retention process of some guanidine based compounds, *J. Liq. Chromatogr. Rel. Technol.*, *35*: 2042–2053, 2012.

85. M. Radulescu, K. Kuca, K. Musilek, and V. David, Structural modifications of dicationic acetylcholinesterase reactivators studied under ion-pairing mechanism in reversed-phase liquid chromatography, *J. Sep. Sci.*, *37*: 3024–3032, 2014.

86. J. Dai and P. W. Carr, Role of ion pairing in anionic additive effects on the separation of cationic drugs in reversed-phase liquid chromatography, *J. Chromatogr. A*, *1072*: 169–184, 2005.

87. M. Radulescu and V. David, Retention study for some neurotransmitters under ion-pairing liquid chromatographic mechanism, *Rev. Roum. Chim.*, *59*: 437–446, 2014.

88. R. LoBrutto and Y. V. Kazakevich, Chaotropic effects in RP-HPLC, in Advances in Chromatography, (E. Grushka and N. Grinberg, eds.), Taylor & Francis Group, *44*: 291–315, 2005.

89. W. H. Pirkle and T. C. Pochapsky, Considerations of chiral recognition relevant to the liquid chromatography separation of enantiomers, *Chem. Rev.*, *89*: 347–362, 1989.

90. I. Paci, I. Szleifer, and M. A. Ratner, Chiral separation: Mechanism modeling in two-dimensional systems, *J. Amer. Chem. Soc.*, *129*: 3545–55, 2007.

91. R. E. Boehm, D. E. Martire, and D. W. Armstrong, Theoretical considerations concerning the separation of enantiomeric solutes by liquid chromatography, *Anal. Chem.*, *60*: 522–528, 1988.

92. S. Ma, H.-W. Tsui, E. Spinelli, C. A. Busacca, E. I. Franses, N.-H. L. Wang, L. Wu et al., Insights into chromatographic enantiomeric separation of allenes oncellulose carbamate stationary phase, *J. Chromatogr. A*, *1362*: 119–128, 2014.

93. H. F. Zou, Y. K. Zhang, and P. C. Lu, Separation mechanism of chiral compounds in chiral stationary phase liquid chromatograph, *Chinese J. Chem.*, *9*: 231–236, 1991.

94. Y. Xiao, T. T. Tan, and S. C. Ng, Enantioseparation of dansyl amino acids by ultra-high pressure liquid chromatography using cationic β-cyclodextrins as chiral additives, *Analyst*, *136*: 1433–1439, 2011.

95. P. Rodríguez-Bonilla, J. M. López-Nicolás, L. Méndez-Cazorla, and F. García-Carmona, Development of a reversed phase high performance liquid chromatography method based on the use of cyclodextrins as mobile phase additives to determine pterostilbene in blueberries, *J. Chromatogr. B*, *879*: 1091–1097, 2011.

96. A. Rocco, A. Maruska, and S. Fanali, Cyclodextrins as a chiral mobile phase additive in nano-liquid chromatography: Comparison of reversed-phase silica monolithic and particulate capillary columns, *Anal. Bioanal. Chem.*, *402*: 2935–2943, 2012.

97. L. Yu, S. Wang, and S. Zeng, Chiral mobile phase additives in HPLC enantioseparations, *Methods Mol. Biol., 970*: 221–231, 2013.

98. S. M. Han, Direct enantiomeric separations by high performance liquid chromatography using cyclodextrins, *Biomed. Chromatogr., 11*: 259–271, 1997.

99. S. Ma, S. Shen, N. Haddad, W. Tang, J. Wang, H. Lee, N. Yee, C. Senanayake, and N. Grinberg, Chromatographic and spectroscopic studies on the chiral recognition of sulfated beta-cyclodextrin as chiral mobile phase additive enantiomeric separation of a chiral amine, *J. Chromatogr. A, 1216*: 1232–1240, 2009.

100. C. Pettersson and G. Schill, Separation of enantiomers in ion-pair chromatographic systems, *J. Liq. Chromatogr., 9*: 269–290, 1986.

101. A. Karlsson and C. Pettersson, Separation of enantiomeric amines and acids using chiral ion-pair chromatography on porous graphitic carbon, *Chirality, 4*: 323–332, 1992.

102. R. Bhushan and S. Tanwar, Direct TLC resolution of atenolol and propranolol into their enantiomers using three different chiral selectors as impregnating reagents, *Biomed. Chromatogr., 22*: 1028–1034, 2008.

103. C. Pettersson, Chromatographic separation of enantiomers of acids with quinine as chiral counter ion, *J. Chromatogr. A, 316*: 553–567, 1984.

104. G. Bazylak and H. Y. Aboul-Enein, Direct separation of labetalol stereoisomers in high performance liquid chromatography system using helically self-assembled chelate as chiral selector in the mobile phase, *J. Liq. Chromatogr. Rel. Technol., 22*: 1171–1192, 1999.

105. W. Hu, P. R. Haddad, K. Hasebe, M. Mori, K. Tanaka, M. Ohno, and N. Kamo, Use of a biomimetic chromatographic stationary phase for study of the interactions occurring between inorganic anions and phosphatidylcholine membranes, *Biophys. J., 83*: 3351–3356, 2002.

106. M. Chrysanthakopoulos, F. Tsopelas, and A. Tsantili-Kakoulidou, Biomimetic chromatography: A useful tool in the drug discovery process, *Advances in Chromatography* (E. Grushka and N. Grinberg, eds.), *51*: 91–125, 2014.

107. F. Tsopelas, M. Ochsenkühn-Petropoulou, and A. Tsantili-Kakoulidou, Void volume markers in reversed-phase and biomimetic liquid chromatography, *J. Chromatogr. A, 1217*: 2847–2854, 2010.

108. H. Luo and Y.-K. Cheng, A comparative study of void volume markers in immobilized-artificial-membrane and reversed-phase liquid chromatography, *J. Chromatogr. A, 1103*: 356–361, 2006.

109. E. S. Gallagher, E. Mansfield, and C. A. Aspinwall, Stabilized phospholipid membranes in chromatography: Toward membrane protein-functionalized stationary phases, *Anal. Bioanal. Chem., 406*: 2223–2229, 2014.

110. D. Vrakas, C. Giaginis, and A. Tsantili-Kakoulidou, Different retention behavior of structurally diverse basic and neutral drugs in immobilized artificial membrane and reversed-phase high performance liquid chromatography: Comparison with octanol-water partitioning, *J. Chromatogr. A, 1116*: 158–164, 2006.

111. D. Vrakas, C. Giaginis, and A. Tsantili-Kakoulidou, Electrostatic interactions and ionization effect in immobilized artificial membrane retention A comparative study with octanol-water partitioning, *J. Chromatogr. A, 1187*: 67–78, 2008.

112. T. Halicioğlu and O. Sinanoğlu, Solvent effects on *cis-trans* azobenzene isomerization: A detailed application of a theory of solvent effects on molecular association, *Ann. N.Y. Acad. Sci., 158*: 308–317, 1974.

113. C. Horvath, W. Melander, and I. Molnar, Solvophobic interactions in liquid chromatography with nonpolar stationary phases, *J. Chromatogr., 125*: 129–156, 1976.

114. O. Sinanoğlu, The C-potential surface for predicting conformations of molecules in solution, *Theor. Chim Acta, 33*: 279–284, 1974.

115. S. C. Moldoveanu and V. David, Dependence of distribution constant in liquid-liquid partition equilibria on van der Waals molecular surface area, *J. Sep. Sci.*, *36*: 2963–2978, 2013.

116. O. Sinanoğlu, Solvent effects on molecular associations. In *Molecular Associations in Biology*; Pullman, B. Ed.; Academic Press, New York, 1968; pp. 427–445.

117. W. J. Lyman, W. F. Reehl, and D. H. Rosenblatt, *Handbook of Chemical Property Estimation Methods*, ACS, Washington, DC, 1990.

118. T. C. Liu, Application of Kihara parameters in conventional force fields, *J. Math. Chem.*, *48*: 363–369, 2010.

119. S. Moldoveanu and V. David, Estimation of phase ratio in reversed-phase high performance liquid chromatography, *J. Chromatogr. A*, *1381*: 194–201, 2015.

120. E. Caiali, V. David, H. Y. Aboul-Enein, and S. C. Moldoveanu, Evaluation of the phase ratio for three C18 high performance liquid chromatographic columns, *J. Chromatogr. A*, *1435*: 85–91, 2016.

121. V. H. Dalvi and P. J. Rossky, Molecular origin of fluorocarbon hydrophobicity, *PNAS*, *107*: 13603–13607, 2010.

122. D. Casoni, J. Petre, V. David, and C. Sarbu, Prediction of pesticides chromatographic lipophilicity from the computational molecular descriptors, *J. Sep. Sci.*, *34*: 247–254, 2011.

123. M. Vitha and P. W. Carr, The chemical interpretation and practice of linear salvation energy relationships in chromatography, *J. Chromatogr. A*, *1126*: 143–194, 2006.

124. M. H. Abraham, H. S. Chadha, and A. J. Leo, Hydrogen bonding: XXXV. Relationship between high-performance liquid chromatography capacity factors and water-octanol partition coefficients, *J. Chromatogr. A*, *685*: 203–211, 1994.

125. M. H. Abraham and W. E. Acree Jr, Descriptors for ions and ion-pairs for use in linear free energy relationships, *J. Chromatogr. A*, *1430*: 2–14, 2016.

126. R. K. Lindsey, J. L. Rafferty, B. L. Eggimann, J. I. Siepmann, and M. R. Schure, Molecular simulation studies of reversed-phase liquid chromatography, *J. Chromatogr. A*, *1287*: 60–82, 2013.

127. J. L. Rafferty, J. I. Siepmann, and M. R. Schure, Understanding the retention mechanism in reversed-phase liquid chromatography: Insights from molecular simulation, in: *Advances in Chromatography* (P. Brown, E. Grushka, eds.), Marcel Dekker, New York, *48*: 1–55, 2010.

128. J. L. Rafferty, L. Zhang, J. I. Siepmann, and M. R. Schure, Retention mechanism in reversed-phase liquid chromatography: A molecular perspective, *Anal. Chem.*, *79*: 6551–6558, 2007.

4 Immobilized Chiral Selectors on Monolithic High-Performance Liquid Chromatography Columns

Ali Fouad and Ashraf Ghanem

CONTENTS

ABBREVIATIONS

AGP	alpha(1)-acid glycoprotein
C18-TMSPAC	*N*-octadecyldimethyl(3-(trimethoxysilyl)propyl)ammonium chloride
CCC	counter current chromatography
CDI	carbonyldiimidazole
CDMPC	cellulose *tris*(3,5-dimethylphenylcarbamate)
CE	capillary electrophoresis
CEC	capillary electrochromatography
CLC	capillary liquid chromatography
CNT	carbon nanotube
CPCE	4-cyanophenyl dicyclohexyl propylene
CP-silica	hybrid chloropropyl-functionalized silica
CSs	chiral selectors
CSPs	chiral stationary phases
DCIB	3,5-dichlorobenzoyl chloride
DCIB-amino acids	3,5-dichlorobenzoyl-amino acids
D-DBTA	dibenzoyl-D-tartaric acid
DETA	diethylenetriamine
DMB	3,5-dimethoxybenzoyl-amino acids
DNB-amino acids	3,5-dinitrobenzoyl-amino acids
DSC	disuccinimidyl carbonate
EDA	ethylene diamine
EDMA	ethylene dimethacrylate
FMOC	9-fluorenylmethoxycarbonyl
GC	gas chromatography
GMA	glycidyl methacrylate
RP	Reversed phase
NP	Normal phase
NAS	N-acryloxysuccinimide
SEM	Scanning electron microscopic
DMSO	dimethyl sulfoxide
XPS	X-ray photoelectron spectroscopy
HEMA	hydroxyethyl methacrylate
HSA	human serum albumin
ID	inner diameter
LC	liquid chromatography
LC-MIM	liquid-crystalline molecularly imprinted monolith

MBQD	O-9-($tert$-butylcarbamoyl)-11-[2-(methacryloyloxy) ethylthio]-10,11-dihydroquinidine
MIP	molecularly imprinted polymer
MMA	methyl methacrylate
p-NB	p-nitrobenzoyl-amino acids

4.1 INTRODUCTION

Enantioselective chromatography has been well documented as a powerful, contemporary, and practical technique for the chiral separation of racemic drugs, food additives, agrochemicals, fragrances, and chiral pollutants [1,2]. Indeed, this technique is at the up front of other methods used to access pure enantiomers, including synthesis from chirality pool, asymmetric synthesis from prochiral substrates, and resolution of racemic mixtures [3]. At the industry level, the enantioselective chromatographic separation of racemic mixtures is considered as the most feasible method compared to the time-consuming and expensive synthetic approaches [4].

Remarkable developments took place in the field of enantioselective chromatography since the first chiral separation of enantiomers using optically active stationary phase in mid-1960s [5]. Following this development, several subclasses have emerged as well-established chromatographic techniques with outstanding applications in chiral separation such as electrochromatography (EC), supercritical fluid chromatography (SFC), counter current chromatography (CCC), gas chromatography (GC), and high-performance liquid chromatography (HPLC) [6]. The latter is the most widely used technique in the enantioselective separations of racemates.

Most enantioselective separations by HPLC are done via direct resolution with a chiral stationary phase (CSP) where the chiral resolving agent is adsorbed, attached, bound, or immobilized to an appropriate support to make a CSP. The enantiomers are resolved by the formation of temporary diastereomeric complexes between the analyte and the CSP. With the development of enantioselective chromatography, many chiral selectors (CSs) such as polysaccharides (Chiracel, Chiralpak, Lux), crown ethers, antibiotics (Chirobiotic T & V) have been introduced and documented as potent chiral stationary phases (CSPs) for the resolution of racemates.

Despite such advances and commercial achievements in CSPs, the stationary phase support represents a crucial point of research investigation. Silica has been recognized as the most popular stationary phase support for HPLC. However, its narrow pH range and thermal instability led to some limitations [7–9]. Several porous adsorbent particles like titanium oxide and zirconia have been introduced as silica substitutes. However, the introduction of porous monolithic material as support for HPLC stationary phases represents an outstanding inspiration in chromatography.

A monolith consists of a single piece of porous solid, which possesses a highly efficient continuous homogenous stationary phase layer (sealed against the wall of a tube/column). The mobile phase has to percolate through the continuous porous bed. Because of their enhanced mass transfer and lower pressure drop, monolithic columns represent a good alternative to particle-packed columns for both capillary electrochromatography (CEC) and HPLC analyses [10–12]. Controlled pore size, increased mass transfer, lower pressure, feasible preparation, and avoided need for end frits are some of the advantages of monolithic columns [13]. Monoliths can be used in both *conventional* HPLC columns (2–4.8 mm i.d.) and nano-HPLC capillaries (up to 500 µm i.d.) [14].

On the other hand, particles-packed stationary phase with reduced particle size (5 µm or less) leads to high pressure, slower flow rates, difficulties with preparation, and the need for end frits. Although ultra high pressure liquid chromatography systems offered an acceptable management of elevated back pressure, its price represents a great limitation [15]. A promising alternative is exemplified by porous monolithic material, which exhibits controllable sizes of the throughpore channels during preparation allowing for a high volume of throughpores, resulting in a low backpressure exhibited by the monolith. Furthermore, monolithic stationary phases show high hydraulic permeability and a dominance of convection over the diffusion mechanism of mass-exchange under dynamic conditions allowing for separations to be carried out at extremely high flow rates [16]. This feature offers higher efficiency columns because they can be longer, and mobile phase can be flowed faster, all while still maintaining an acceptable head pressure. The advantages of monolithic stationary phases (SPs) over particle-packed equivalents are highlighted in Table 4.1.

TABLE 4.1
Advantages and Disadvantages of Using Monoliths Compared to Alternative Particle-Packed Phases for Separation

	Monolithic SP	Particle Packed SP
Advantages	Low cost, relative ease of *in situ* preparation	High surface area
	Mechanically robust, no void volumes formed with conventional LC flow rates	Higher column efficiencies
	Can control the porous properties by varying starting materials for various sized target molecules	Small particle sizes and high-operating pressures
	High permeability and better convection over the diffusion mechanism of mass exchange under dynamic conditions allowing the analysis to be carried out at extremely high flow rates	Diverse column chemistries and HPLC column dimensions
	Flexible synthesis enables monoliths to be modified by ion-exchange, affinity, chiral, mixed-mode, restricted access, hydrophobic, and hydrophilic to tailor the stationary phase for different analytes	Validated applications and assays
	Can be molded into any shape (capillary, column, micropipette tips, microfluidic channel on a chip)	
	Relatively biocompatible as open porous structure	
	Can withstand more extreme conditions (e.g., pH working range from 2 to 10)	

(*Continued*)

TABLE 4.1 (*Continued*)

Advantages and Disadvantages of Using Monoliths Compared to Alternative Particle-Packed Phases for Separation

	Monolithic SP	Particle Packed SP
Disadvantages	Lower surface area and binding capacity	Higher backpressure (slow diffusional mass transfer)
	Lower column efficiencies and HPLC column to column reproducibility	Needs a pump to generate high pressure for the sample to pass through and the problem of all particulate media is their inability to completely fill the available space. The channeling between particles reduces extraction efficiencies and can adversely affect flow rates
	Narrow column chemistries and column dimensions commercially available	Fabrication requires a high skill due to small i.d. particles used
	Limited use in routine analysis due to the limitation of commercial suppliers	Need for frits, undesired interactions may occur, and can make problems
		Narrow pH stability range

4.2 MONOLITHIC STATIONARY PHASES

Three main classes of monolithic stationary phases have been reported, namely (1) organic polymer-based, (2) silica-based, and (3) hybrid monolithic stationary phases [17].

4.2.1 ORGANIC POLYMER MONOLITHIC STATIONARY PHASES

Polymerization has provided an opportunity in making a more homogeneous polymer, which is favorable for polymer-based monolithic column fabrication [18]. The synthesis of organic polymer monolithic stationary phases involves an *in situ* polymerization of monomeric precursors (e.g., butyl acrylate) and cross-linker in a porogenic solvent (e.g., 1,3-butanediol diacrylate), in presence of an initiator (e.g., 2,2′-azobis (2-methylpropionitrile [AIBN]) (Figure 4.1) [19,20]. Pore size and structure depend on the type of porogenic solvent [21]. For monolithic media, poor solvents of the chosen monomers resulted in the formation of the macro-throughpores sought for liquid flow. To introduce chirality, CSs might be mixed with the prepolymerization slurry mixture to be bonded or immobilized on or functionalize the monolith.

The prepolymerization surface modification with a suitable precursor leads to increased stability of the monolith and affords greater adherence to the confining wall. Theoretically, any monomer can be used to form a monolith. Consequently, monoliths with diverse chemistries could be prepared as hydrophobic (using styrene or butyl methacrylate), hydrophilic (using acrylamide or 2-hydroxyethylmethacrylate),

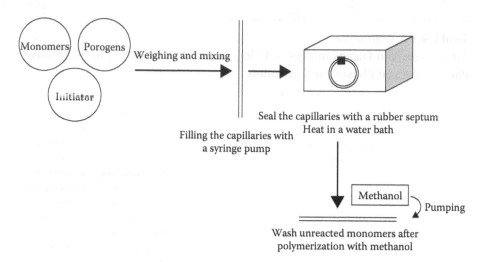

FIGURE 4.1 Schematic diagram showing the *in situ* preparation of polymer monolithic capillary columns.

ionizable (using 2-acrylamido-2-methyl-1-propanesulfonic acid), reactive (using glycidyl methacrylate or 2-vinyl-4,4-dimethylazalactone), and chiral (using (R)-acryloyloxy-β-β-dimethyl-γ-butyrolactone) [22].

4.2.2 Silica-Based Monolithic Stationary Phases

Silica is the most popular inorganic material used in liquid chromatography (LC) and their columns are commercially available in different diameters from 4.6 mm ID down to 100 μm ID. The preparation of a silica monolith employs the classical sol-gel process, which started firstly by the hydrolysis of a silane precursor (e.g., tetraalkoxy silanes mostly tetramethoxysilane [TMOS], tetraethoxysilane [TEOS], or methyltrimethoxysilane [MTMS]) in the presence of a porogen (e.g., polyethylene oxide, polyethylene glycol [PEG], or polyacrylic acid). During the condensation of silica (gelation) in the presence of the porogen, a phase separation takes place between silica and porogen system and water, forming a bicontinuous network of macropores and a micrometer-sized silica skeleton, then the polycondensation product can precipitate as particles or a monolithic mass with large throughpores. The size and distribution of the throughpores are determined by silane precursor(s) and the porogens concentration. Secondly, mesopores are tailored inside the silica skeleton by a dissolution-reprecipitation process in a moderate alkaline medium (using urea or ammonia) at temperature close to 120°C (aging step). Finally, sintering step (drying and calcination) is carried out at 330°C or higher in order to remove the organic moieties from the monolith (Figure 4.2) [23,24]. The monolithic columns are then ready for characterization or further modification to introduce different surface chemistries [25–27].

As described earlier, the size of the meso- and macropores can be achieved independently as they are generated via two sequential steps. As a consequence, silica monoliths allow fast and highly effective separation of small or low molecular weight

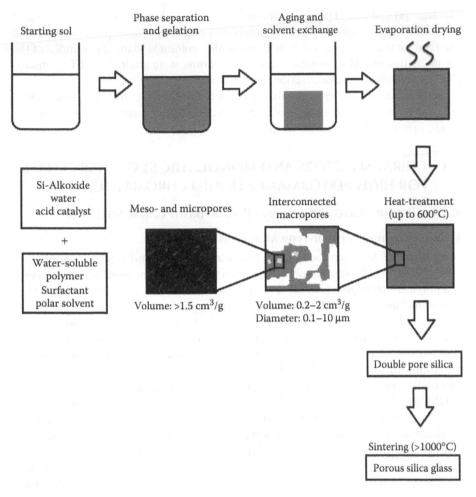

FIGURE 4.2 Schematic representation showing the process of the *in situ* preparation of silica-based monolithic stationary phases (MSPs). (From Nakanishi, K. et al., *J. Sol-Gel Sci. Technol.*, 13, 163–169, 1998.)

compounds due to its easy penetration of the adsorption sites and fast diffusion (through macropores). The efficiency is lower for high molecular weight compounds because the diffusion of large molecules into the mesopores is much slower. Silica monoliths have rigid inorganic skeleton, thus it displays limited swelling by solvents, which leads to similar pore morphology of the silica support in dry and wet state [28].

4.2.3 Hybrid Monolithic Stationary Phases

Organic-silica hybrid monolithic columns have drawn more attention due to the ease of preparation and good mechanical and pH stability in recent years [29]. Varieties of hybrid monolithic capillary columns have been prepared based on many established synthetic methodologies. The sol–gel process is well recognized in the fabrication of hybrid monolithic columns, which can be mainly categorized as one-step, acid/base two-step procedures. The new approaches such as the *one-pot* and nanoscaled inorganic–organic hybrid reagent of polyhedral oligomeric silsesquioxane have also emerged for the preparation of hybrid monolithic columns. Many applications of the organic-silica hybrid monolithic capillary columns were studied for CEC, micro-HPLC, solid-phase microextraction, and enzymatic reactor [30].

In this chapter, various immobilized, bonded, and functionalized CSs onto mono-lithic columns and capillaries have been studied within enantioselective conven-tional, micro- and nano-HPLC formats.

4.3 CHIRAL SELECTORS AND MONOLITHIC STATIONARY PHASES FOR HIGH-PERFORMANCE LIQUID CHROMATOGRAPHY

4.3.1 PROTEIN, GLYCOPROTEIN, AND PEPTIDE-BASED CHIRAL SELECTORS

4.3.1.1 Organic Polymer-Type Monoliths

Several studies have reported on the preparation of protein-based polymer mono-liths as CSs for chromatographic separation using different monomers, porogens, and polymerization protocols [31–38]. Among these studies, poly(GMA-*co*-EDMA) monoliths have been used extensively for postmodification approaches due to the presence of reactive epoxy groups susceptible to reaction with various nucleo-philes [39–42]. Affinity monoliths where proteins can be covalently immobilized onto monolithic surface were prepared via a simple nucleophilic reaction of amine-containing ligands with poly(GMA-*co*-EDMA) monoliths. Examples of covalent immobilization methods used to attach proteins and other amine-containing agents to GMA/EDMA monoliths include the (a) epoxy method, (b) Schiff base method, (c) carbonyldiimidazole (CDI) method, (d) disuccinimidyl carbonates (DSC) method, (e) glutaraldehyde method by ethylenediamine or hexanediamine, (f) the hydrazide method, and (g) cyanogen bromide (CNBr) method (Figure 4.3) [39,43–54].

Human serum albumin (HSA) was immobilized using the Schiff base method and the epoxy method. The highest protein content was obtained by the Schiff base technique, whereas the epoxy technique gave the lowest protein content. In terms of separation, the best resolution of (*R/S*)-warfarin and D/L-tryptophan was achieved by the Schiff base-immobilized column. All methods gave similar activities for HSA in its binding to (*R*)- and (*S*)-warfarin; however, some differences were noted in the activity of the immobilized HSA for D/L-tryptophan [43].

Yao et al. reported on the immobilization of HSA on poly(GMA-*co*-EDMA) in high-performance affinity chromatography using the glutaraldehyde method by ethylene diamine (EDA) and epoxy method (Figure 4.3). The monoliths were successfully adopted for the chiral separation of *d,l*-amino acids (AAs) and the quan-tification of D-tryptophan in urine samples. The polymerization conditions and the techniques for immobilization of HSA on GMA/EDMA and GMA/TRIM mono-liths were thoroughly studied, with particular emphasis being given to the effects of changing the composition of the porogen [55].

In a further study, both the Schiff base and epoxy immobilization methods were used in similar monoliths. A 2.6–2.7-fold increase in HSA content was obtained in the final monoliths when compared to similar HSA monoliths prepared by other

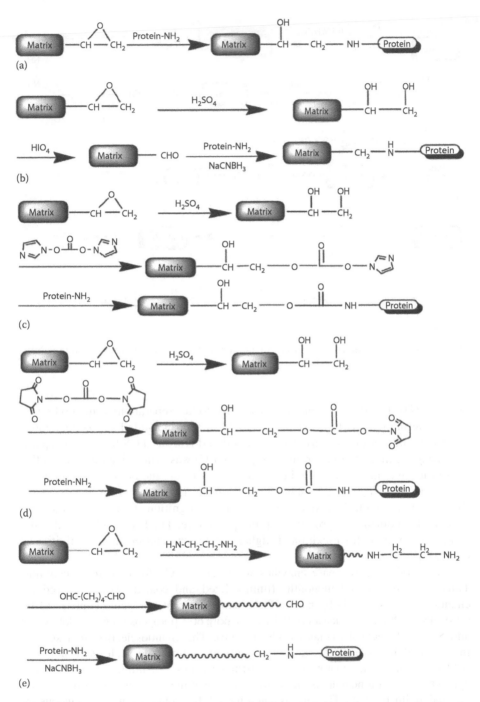

FIGURE 4.3 (a) The epoxy method, (b) the Schiff base method, (c) the carbonyldiimidazole (CDI) method, (d) disuccinimidyl carbonates (DSC) methods, (e) the glutaraldehyde method by ethylenediamine. *(Continued)*

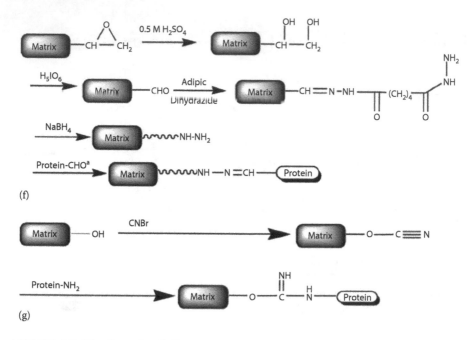

(f)

(g)

FIGURE 4.3 (Continued) (f) The hydrazide method, and (g) cyanogen bromide (CNBr) method.

methods [43]. The increased protein content led to fast separations with good retention and resolution for chiral agents such as *R/S*-warfarin and D/L-tryptophan [56].

Monolithic stationary phase with immobilized HSA as a CS for enantioseparation using capillary liquid chromatography (CLC) was synthesized in fused-silica capillaries in a modified method using additives during the protein allylation and polymerization steps. Acetyl salicylic acid and L-tryptophan were used as additives to interact with the active sites for chiral recognition (Figure 4.4), thus preserving them during the protein-coupling procedure. The beds synthesized using L-tryptophan as additive exhibited higher resolutions and enantioselectivities in CLC [57].

On the other hand, lipase enzymes were used as CS for monolithic columns. Thus, two lipase-based monoliths (immobilized and coated) were prepared for enantioselective analysis by nano-HPLC using different porogenic content (61.1% porogens [54.2% formamide and 6.9% 1-propanol]; 60% porogens [51.6% cyclohexanol and 8.4% 1-dodecanol]) (Figures 4.5 and 4.6). The enantioselective analysis was investigated for many sets of pharmaceutical compounds such as β-blockers, α-blockers, anti-inflammatory drugs, antifungal drugs, dopamine antagonists, norepinephrine dopamine reuptake inhibitor, catecholamines, sedative hypnotics, diuretics, and antihistamines. The encapsulated monolithic column was advantageous for several categories of chiral drugs [20].

Allyl cellulase was entrapped in (encapsulated) and covalently immobilized on epoxy-activated monolithic CSPs. The prepared beds were used for the

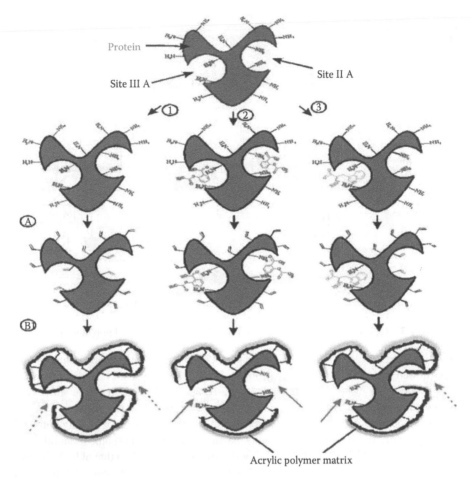

Acrylic polymer matrix

FIGURE 4.4 Schematic drawing of the HSA immobilization procedure without using an additive (branch 1), using acetylsalicylic acid (branch 2), and using L-tryptophan (branch 3) as an additive. Step A—allylation of the protein and Step B—continuous bed synthesis.

enantioselective separation of some β-blockers. Of particular interest, practolol was separated rapidly and with high resolution [58].

4.3.1.2 Silica-Based Monoliths

Silica monoliths have a high level of immobilized protein attachment due to its potential advantages such as the high surface area and their ability to use the same immobilization methods that were employed when attaching different affinity ligands to silica particles [59]. In silica monoliths, the same immobilization techniques have been used such as epoxy method, Schiff base, but epoxy method tends to give lower protein coverage for HSA and lower activities than other amine-based coupling methods [43,60].

The immobilization protocol for the covalent linkage of HSA to silica monolithic support is reported and involves the use of periodic acid to oxidize a diol-derived silica

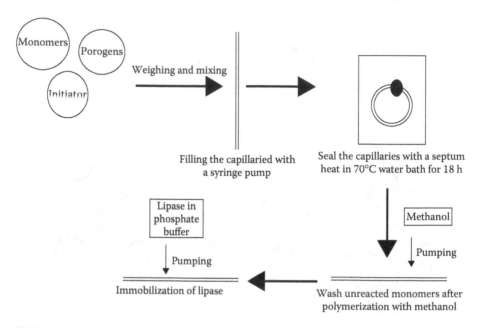

FIGURE 4.5 General approach for preparation of lipase-immobilized polymer monolith.

monolith yielding aldehyde groups on its surface. The monolith aldehyde groups are then allowed to react with the amine groups to form a Schiff base. The reduction of the C=N group using cyanoborohydride then takes place to generate the more stable secondary amine linkage [61]. The enantioseparation of tryptophan, ibuprofen, and warfarin was described, and the chromatographic results were compared with data obtained for the same protein when coupled to silica particles or to a rigid organic monolith. The surface coverage is 1.3–2.2 times higher for the silica monolith than that of both the particulate silica and the rigid organic monolith which were nearly the same, but the HSA silica monolith gave higher comparable resolutions and efficiencies [61].

In the further two methods, HSA has been immobilized via its sulfhydryl groups using of maleimide-activated silica (the SMCC method) or iodoacetyl-activated silica (the SIA method). The resulting supports were tested for use in chromatography using HSA as a model protein. The studies indicated that iodoacetamide-modified HSA had a high selectivity for sulfhydryl groups, which accounted for the coupling of 77%–81% of this protein to maleimide- or iodoacetyl-activated silica. Total protein content, specific activity, nonspecific binding, binding capacity, stability, and chiral selectivity for several analytes were evaluated. Maleimide-activated monolithic silica HSA columns afforded the best results for the aforementioned properties when compared to HSA immobilized by Schiff base method. The supports developed in this work offer the potential of giving greater site-selective immobilization and ligand activity than amine-based coupling methods. These features render such supports attractive in the development of protein columns for the study of biological interactions and chiral separations [62].

In a related work, trypsin was immobilized on an epoxy-modified silica monolithic support with a single reaction step for online protein digestion and peptide

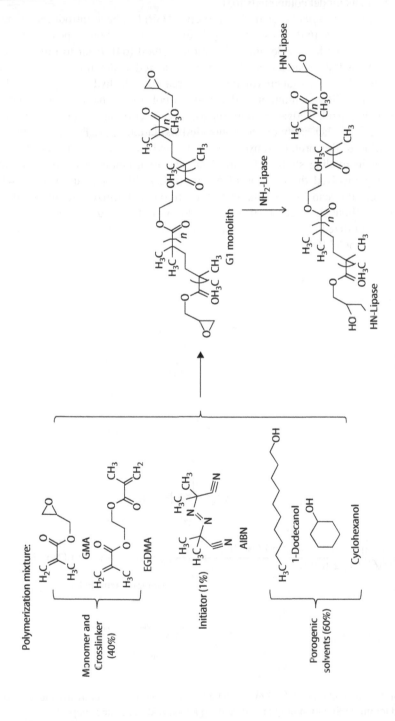

FIGURE 4.6 Schematic diagram showing the polymer backbone of lipase-based monolith.

analysis. The efficacy of the method for tryptic mapping is reported for somatostatin and myoglobin as model compounds [63].

In another report, *alpha*(1)-acid glycoprotein (AGP) has been introduced as new CS by Xuan and Hage [64]. Thus, AGP has been immobilized onto epoxy-activated supports through the hydrazide immobilization method [64]. Prior to immobilization, the carbohydrate residues on AGP were oxidized under mild conditions to generate aldehyde groups (approximately, five reactive aldehyde groups per AGP molecule) [65]. AGP was immobilized via its carbohydrate chains after periodate oxidation to hydrazide-activated supports for use in chromatographic or biosensor technology. Further study on the controlled oxidation of AGP followed by the immobilization of this protein to hydrazide-activated silica was carried out for use in drug–protein binding studies (Figure 4.7). The final conditions chosen for the oxidation step involved the reaction of 5.0 mg/mL AGP at 4°C and pH 7.0 with 5–20 mM periodic acid for 10 min. These conditions maximized the immobilization of AGP without significantly affecting its activity. The results revealed that the resulting immobilized AGP afforded good qualitative and quantitative analysis for *R*- and *S*-propranolol [64].

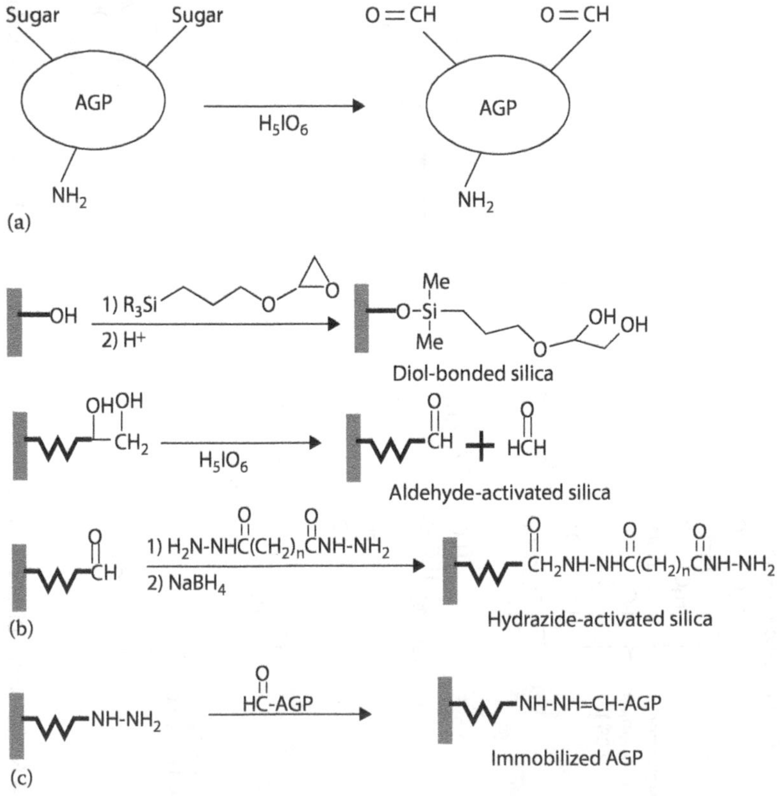

FIGURE 4.7   (a) Oxidation of AGP by periodate acid, (b) preparation of hydrazide-activated silica, and (c) immobilization of oxidized AGP to a hydrazide activated support.

Mallik et al. [66] reported the AGP CS immobilized on affinity silica mono-liths used for the HPLC enantiomeric separation of warfarin and propranolol. The chromatographic results were compared with data obtained from AGP-immobilized poly(GMA-*co*-EDMA) monolith or AGP-immobilized silica particles. The surface coverage of AGP in the silica monolith was about 61% higher than that observed for the organic polymer-based monolith and about 18% higher than that obtained with silica particles.

In different context, cyclic hexapeptide (CS) molecule as a CS was immobilized onto the surface of a monolithic support for the HPLC enantiomeric separation of a series of dansyl-amino acids, namely dansyl-alanine (dan-ala), dansyl-valine (dan-val), dansyl-norvaline (dan-nor), dansyl-leucine (dan-leu), dansyl-phenylalanine (dan-phe), and dansyl-tryptophane (dan-try) (Figure 4.8). Arylalkanoic acids including etodolac

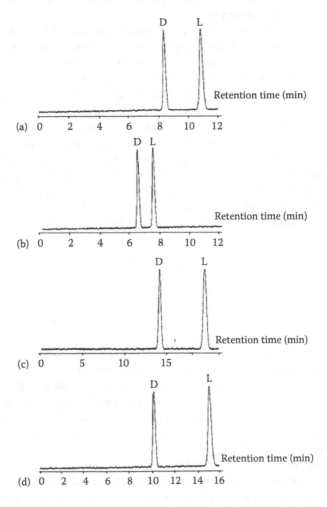

FIGURE 4.8 Chromatographic enantioseparation of (a) dan-nor, (b) dan-try, (c) flu, and (d) nap. Mobile phase: phosphate buffer (M/15) (pH 5.6); flow rate: 1 mL/min; $T = 25°C$.

(eto), flobuphen (flo), ibuprofen (ibu), flurbiprophen (flu), naproxen (nap), and sulindac (sul) were also tested as racemic standards using HPLC (Figure 4.8). The enantiosepa-rations are based on the chiral recognition properties of the cyclic hexapeptide mol-ecule and the unique properties concerning the flow behavior of silica monoliths [67].

Furthermore, new monolithic silica columns were bonded to two polyproline-derived CSs, which considered constituting a new class of peptide-based CSs. The enantioseparation ability of the resulting chiral monolithic columns was evaluated with a series of structurally different racemic test compounds. When compared to analogous bead-based CSPs, higher enantioseparation and broader application domain were observed for polyproline-derived monolithic columns. Monolithic silica matrices allowed the introduction of higher amounts of polyproline-based CSs onto the CSP than their particle-based counterparts. This increased CS density leads to greater separation factors, higher retention times, and a broader application domain for monolithic columns [68].

Additionally, the structural difference between the studied CSs also influences their bonding to the silica gel surface, thus affecting retention time for the two col-umns tested. The higher surface coverage of monolithic columns by CS results in an increased loadability, which resulted in improved accessibility of the CS to the ana-lyte when bonded to a monolithic matrix. This property renders polyproline-derived monolithic CSPs suitable for preparative purposes. Considering analytical applica-tions, it is worth noting that the increase in flow rate permits the reduction of analysis time with only a minor effect on resolution. This feature renders this kind of mono-lithic CSPs suitable candidates for the application to high-throughput approaches [68].

An oligopeptide of proline derivative (octaproline)-based chiral monolithic col-umn, having 3,5-dimethylphenylcarbamate residues on each proline unit, was cova-lently bonded to a monolithic silica rod for an improved HPLC performance. The new columns showed higher enantioselectivity, broader application domain, and higher loading capacity than the bead-based counterpart. These improved features could be mainly attributed to the increased amount of CS contained into the mono-lithic column. The easier accessibility of the analyte to the CS was produced by the format of the chromatographic matrix, as the matrix format provides a more reactive surface onto which the CS was more easily attached. The monolithic structure of the matrix may supply an additional factor by making the CS more available to the analytes contributing to the observed improvements [69].

In other prospect, a polymethacrylamide backbone was prepared by suspend-ing a CS (*l*-phenylalaninamide, *l*-alaninamide and *l*-prolinamide)-modified silica in the mixture of the monomers followed by *in situ* polymerization in the capillary to form particle-loaded monoliths. The separation was performed following the ligand exchange using electrolytes containing Cu (II) ions. Particle-loaded monoliths were used for capillary-LC and CEC, where dansyl-amino acid enantiomers and hydroxy acid enantiomers were resolved by micro-HPLC (Figure 4.9) [70,71].

Furthermore, new chiral monolithic columns were prepared by pumping a solution of *N*-decyl-L-4-hydroxyproline, *N*-hexadecyl-L-4-hydroxyproline, or *N*-2-hydroxydodecyl-L-4-hydroxyproline followed by loading of copper (II) ions through a commercially available monolithic reverse-phase column. The columns were applied to the chiral separation of amino acids, diastereomeric dipeptides,

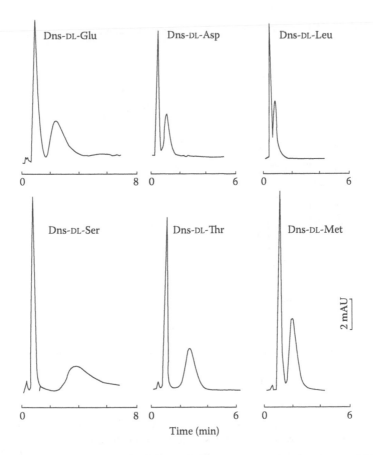

FIGURE 4.9 Representative chromatograms for the enatioseparations of dansyl-amino acids using L-phenylalaninamide-modified monolithic columns.

glycyl dipeptides, and tripeptides. Ultrafast separations in the range of seconds were achieved using high flow rates (Figure 4.10). The CSs were fixed with their hydrophobic moieties on the monolith by hydrophobic interactions, and the separation was based on the formation of ternary mixed complexes between the selector and the analyte [72,73].

Similar to what was reported in polymer-based monolith, enzyme (penicillin G acylase [PGA]) has been used as CS for monolithic silica column. The column was prepared via *in situ* modification process using epoxy silanes on which PGA was immobilized. The chemically modified monolithic silica column was characterized in terms of epoxy groups attached to the surface. A higher immobilization yield was obtained when compared to the conventional epoxy-silica material. The newly developed chiral stationary phase was successfully used for the enantioseparations combining the well-known chiral recognition properties of PGA and the unique properties of the flow behavior of silica monoliths. A PGA monolithic column can operate at high flow rate without a significant loss in enantioselectivity, and faster

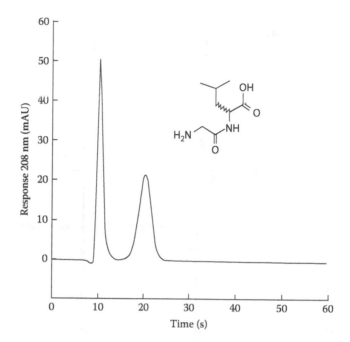

FIGURE 4.10 Chromatogram of the rapid chiral separation of Gly-Leu. Stationary phase: RP-18 monolithic column coated with *N*-decyl-L-4-hydroxyproline. Mobile phase: 0.1 mM Cu(II)sulfate. Injection: 1 μL; flow rate: 8 mL/min.

enantioseparations can be achieved using the prepared supports in high-throughput separations. Fourteen racemic analytes were used for the comparative study [74].

Inspired by the promising separations using conventional chiral PGA-based monolith columns, Gotti et al. developed capillary monolithic columns using PGA as CS. The chiral ability of PGA after immobilization on monolithic support via epoxy technique achieved very fast enantioselective separation of arylpropionic acids under CLC conditions using a short separation column connected to a transparent open capillary section allowing for high detection sensitivity at 200 nm. The new capillary column was used in the determination the (*S*)-ketoprofen in pharmaceutical samples [75].

4.3.2 POLYSACCHARIDE-BASED CHIRAL SELECTORS

4.3.2.1 Silica-Based Monoliths

Covalent immobilization of 3,5-dimethylphenylcarbamate derivative of cellulose via an epoxide moiety was performed on native silica monoliths cladded in a classical 50 mm × 4.6 mm polyether ketone (HPLC) column [76]. The column obtained by this technique has the high enantiomer-resolving ability of the polysaccharide derivative with the favorable dynamic properties of monolithic HPLC columns [76]. This column can be used in with nonstandard mobile phases, which are incompatible with coated-type polysaccharide columns due to the solubility of CS.

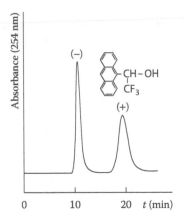

FIGURE 4.11 Enantioseparation of 2,2,2-trifluoro-1-(9-anthryl)ethanol on monolithic silica after covalent immobilization of 16.7% (w/w) of cellulose 3,5-dimethylphenylcarbamate derivative.

The chiral modification of the silica monolith includes reaction with γ-glycidoxyp ropyltrimethoxysilane, modification with the polysaccharide derivative, and then treatment with 3,5-dimethylphenylisocyanate to convert the hydroxyl groups of cellulose into carbamate moieties. The enantioseparation results indicated that the monolithic column still retains the adequate overall enantiomer-resolving ability (Figures 4.9 and 4.11) [76].

In 2006, comparative study between coated and immobilized 3,5-disubstituted phenylcarbamate derivatives of cellulose and amylose onto a native silica monolith was performed. The effects of polysaccharide and substituents as well as of covalent immobilization of polysaccharide derivative on chromatographic performance of capillary columns were studied. The capillary columns obtained using immobilization techniques were robust in all solvents commonly used in LC and exhibit promising enantiomer-resolving ability. Increasing the amount of covalently attached polysaccharide derivative led to improvement of separation, whereas a decrease in the separation efficiency with increasing amount of immobilized CS has been noticed (Figure 4.12) [77].

Further comparative study with the goal to correlate the conversion between conventional and capillary chiral column has been performed. Amylose *tris*-(3-chlorophenylcarbamate) was immobilized on 5 μm or 3 μm silica particles in both conventional and capillary columns for HPLC analysis of a set of different classes of racemic pharmaceuticals, namely anti-inflammatory drugs, antifungal drugs, β-blockers, α-blockers, dopamine antagonists, norepinephrine-dopamine reuptake inhibitors, catecholamines, antihistamines, sedative hypnotics, diuretics, anticancer drugs, flavonoids, and antiarrhythmic drugs (Figure 4.13). Under normal chromatographic condition, baseline separation was achieved for most of analytes on the conventional column and for three analytes on the capillary counterpart. It has been concluded that parameters such as the content and conformation of the CS in both column formats should be considered, and the prediction of

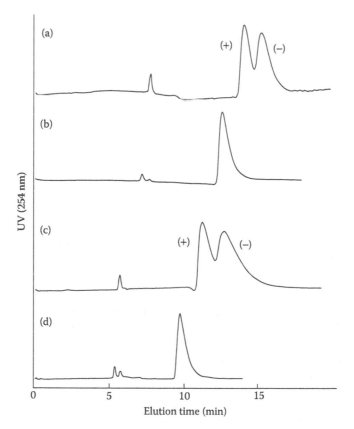

FIGURE 4.12 Chromatograms of *trans*-cyclopropandicarboxylic acid dianilide after the first and the second coating (a and c) and covalent immobilization (b and d) steps of AMDMPC.

column performance when miniaturized cannot be based on the performance of its conventional antipode [78].

In recent study, two CSs namely amylose 3,5-dimethylphenylcarbamate (ADMPC) and cellulose 3,5-dichlorophenylcarbamate (CDCPC) were covalently attached to silica particles of 5 μm in stainless steel columns (4.6 × 250 mm). The two columns were tested for their enantiomer separation ability toward flavanone and its derivatives using different mobile phases as methanol, ethanol, 2-propanol, and acetonitrile modified with 0.1% formic acid. The study revealed that the various column chemistry and mobile phases are quite complementary to each other based on the enantiomeric separation of flavanone and its chiral derivatives [79].

4.3.2.2 Hybrid-Type Monoliths
A chiral hybrid monolithic capillary column was developed by *in situ* coating cellulose *tris*(3,5-dimethylphenylcarbamate) (CDMPC), a polysaccharide derivative onto the hybrid CP-silica monolith, which was modified with diethylenetriamine

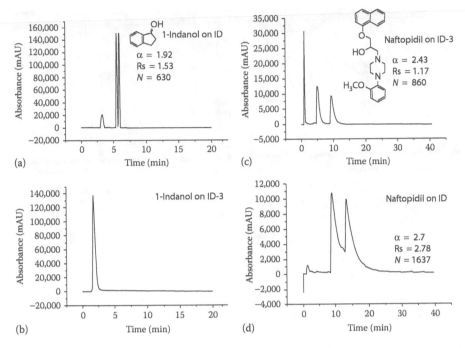

FIGURE 4.13 Enantioselective separations of racemic 1-indanol on the conventional Chiralpak ID (a) vs. the capillary Chiralpak ID-3 (b) and racemic naftopidil on the capillary (c) vs. the conventional Chiralpak ID (d). Condition (a, b): mobile phase: *n*-hexane/2-propanol 90:10 v/v, UV: 254 nm, flow rate: 1 mL/min and 10 µL/min, respectively. Condition (c, d): mobile phase: *n*-hexane/2-propanol 90:10 v/v, UV: 219 nm, flow rate: 50 µL/min and 1 mL/min, respectively.

(DETA) to form NH_2 functionalities via nucleophilic substitution reaction prior to coating. The resulting monolithic capillary was applied for enantioseparation in CLC. The hybrid CP-silica monolith was prepared in a 40 cm capillary. The influences of PEG and urea content as well as polycondensation temperature on the column morphology were further carefully examined. A transparent gel-like monolith was formed in a capillary when PEG was less than 300 mg in the mixture at any temperature. The influence of urea dosage in precondensation mixture on monolith formation and morphology was studied. It was observed that the monolith gradually occupied the whole capillary as the amount of urea was increased to 800 mg. Urea concentrations at 1100, 1200, 1300, and 1400 mg were characterized with SEM, and the results are shown in Figure 4.14. The condensation temperature had an effect on the formation of the CP-silica hybrid monolith. Thus, a series of monoliths were fabricated at different condensation temperatures. The monolith fabricated at 70°C exhibited higher backpressure during the permeability measurement. A variety of racemates were successfully resolved on CP-silica hybrid monolithic column coated with 60 mg/mL CDMPC in CLC with RP and NP modes [80].

FIGURE 4.14   SEM images of the hybrid CP-silica monoliths prepared with different content of urea. (a) 1100 mg, (b) 1200 mg, (c) 1300 mg, and (d) 1400 mg. Magnification: 5000×.

4.3.3   OLIGOSACCHARIDE-BASED CHIRAL SELECTORS

4.3.3.1   Organic Polymer-Type Monoliths

Following the remarkable results of *beta*-cyclodextrin (β-CD) in conventional enantioselective HPLC, a β-CD monomer, namely allyl-hydroxypropyl-β-CD was *in situ* copolymerized with methacrylamide, *N*-isopropylacrylamide, and 1,4-bis(acryloyl) piperazine in the presence of aqueous buffered solution. The polymerization mixture was optimized, and the prepared capillaries were used for the enantioselective separation of nomifensine, praziquantel, 5-methyl-5-phenylhydantoin, and alprenolol using nano-LC [81]. The difficulty to optimize the morphology of monoliths as well as their characterization was the major drawback of the *in situ* copolymerization. Therefore, preoptimized monolithic backbones represents an alternative approach to the *in situ* copolymerization [82].

Further β-CD-immobilized monolithic column has been prepared via click chemistry. β-CD-grafted monolithic stationary phase was prepared in three steps involving successive photopolymerization of a mixture of NAS and EDMA dissolved in toluene as a porogenic agent, nucleophilic substitution of succinimide moieties with propargylamine, and azide-alkyne cycloaddition with mono-(6-azido-6-deoxy)-β-CD. Good enantioseparations of the test chiral compound flavanone enantiomers were achieved under reversed-phase conditions using nano-liquid chromatography

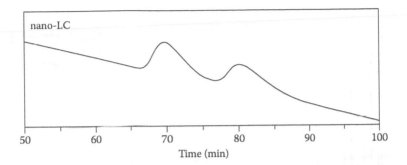

FIGURE 4.15 Resolutions of flavanone enantiomers by nano-LC with methanol-borate buffer (5 mmol.L–1 pH 8.2) 40:60 v/v and detection at 214 nm. Conditions: columns length: 31 cm, flow rate: 60 nL min1, injections: 4 nL. (From Guerrouache, M. et al., *Macromol. Rapid Commun.*, 30, 109–111, 2009.)

(nano-LC) technique. The results demonstrate the potentiality and usefulness of click chemistry in the preparation of β-CD containing chiral organic polymer (Figure 4.15) [83].

In another report, the β-CD was immobilized onto the monolithic column as a CS to improve its biocompatibility. The column stationary phase was prepared by reverse atom transfer radical polymerization (RATRP) with excellent controllability of molecular weight and polydispersity. This monolithic column was used as chiral restricted access stationary phase for the chiral analysis of enantiomers in biological samples. The results indicated that these multifunctional columns can be applied in the determination of chiral drugs in plasma with direct HPLC injection [84].

To investigate the effect of the linking spacer on the enantioseparation ability of β-cyclodextrin (β-CD) polymeric monoliths, three monolithic columns were functionalized with different amino linking spacers, namely mono-6-amino-6-deoxy-β-CD, mono-6-ethylenediamine-6-deoxy-β-CD, and mono-6-hexamethylenediamine-6-deoxy-β-CD affording poly(GMA-NH$_2$-β-CD-*co*-EDMA) (Column A), poly(GMA-EDA-β-CD-*co*-EDMA) (Column B), and poly(GMA-HDA-β-CD-*co*-EDMA) (Column C), respectively. The effects of the linking spacer length on the physicochemical properties including morphology, β-CD density, permeability, and hydrophobicity were thoroughly compared, and no obvious difference could be detected. Furthermore, the enantioseparation ability of these monolithic columns toward 14 chiral acidic compounds was evaluated (Figure 4.16). A triazole linker was compared to the ethylenediamine linker containing monolith. The triazole linker containing β-CD-functionalized column exhibited much lower enantioselectivity, which might be attributed to its lower flexibility. It is worthy to mention that the length of the spacer was greatly affecting the enantioselectivity of β-CD monolithic columns [85].

As previously revealed, the chemical modification of the monolith is limited to the amount of the modifier available in the column, thus producing reduced monolith ligand density. To overcome this limitation, newly prepared

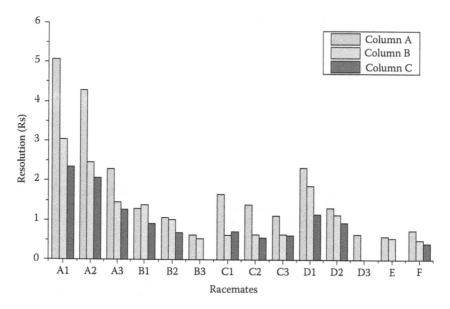

FIGURE 4.16 Comparison of enantioresolution for 14 chiral acidic compounds (A1-F) on three monolithic columns. Conditions: mobile phase: (a) ACN/0.2%TEAAc, pH 4.2 (70/30,v/v) for all tested analytes except D3; (b) ACN/ 0.2%TEAAc, pH 4.2(30/70,v/v) for D3 only; UV detection wavelength: 214 nm; flow rate: 800 nL/min; injection volume: 20 nL. (From Guo, J. et al., *Talanta*, 152, 259–268, 2016.)

ethylenediamine-β-cyclodextrin-functionalized poly(GMA-*co*-EDMA) monoliths were prepared. The XPS elemental analysis used for the characterization showed the appearance of the nitrogen element for the EDA-β-CD-modified monolith. Compared with the blank monolith, the carbon content decreased and the oxygen content increased significantly for the EDA-β-CD-modified monolith used for the chiral separation of racemic ibuprofen by HPLC [86].

Ghanem et al. reported new β-CD-functionalized polymer monoliths prepared via the copolymerization of β-CD-methacrylate and EDMA in different ratios *in situ* in fused silica capillaries (I.D. 150 μm). The monomer 2,3,6-*tris*(phenylcarbamoyl)-β-CD-6-methacrylate was prepared as illustrated in Figure 4.17 [87].

Three polymer-based monolithic capillary columns that were prepared via *in situ* copolymerization of binary monomer mixtures consisted of β-CD-derived functional monomer and EDMA as a cross-linker along with three porogens, namely 1-propanol, 1,4-butanediol, and water in the presence of AIBN (Figure 4.18). The monolithic capillary columns were efficient for the chiral separation of different classes of pharmaceuticals, and the approach could be expanded to other classes of CSs (the separation α and resolution Rs factors for the baseline resolved compounds ranged from α = 1.39–1.89 and Rs = 1.56–2.62, respectively).

The same group recently extended this work to include the preparation of trimethylated-β-cyclodextrin (TM-β-CD) encapsulated in poly(GMA-*co*-EDMA) monolith capillary column using the one-pot *in situ* copolymerization procedure

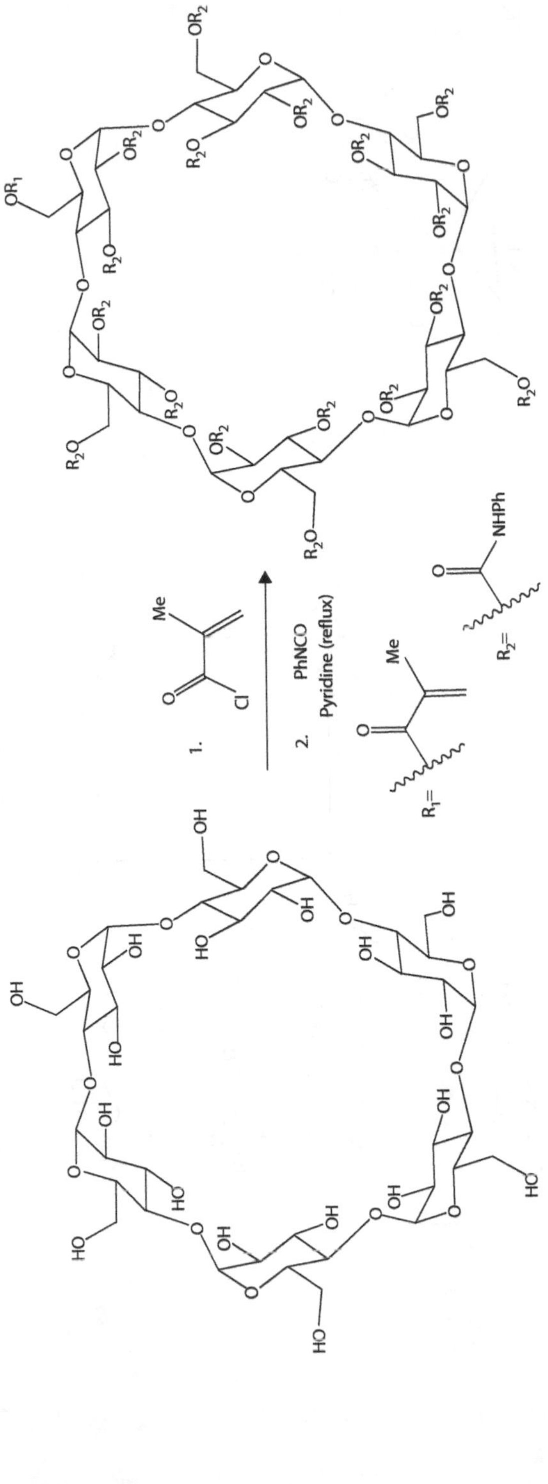

FIGURE 4.17 Schematic illustration for the preparation of 2,3,6-*tris*(phenylcarbamoyl)-β-CD-6-methacrylate chiral selector.

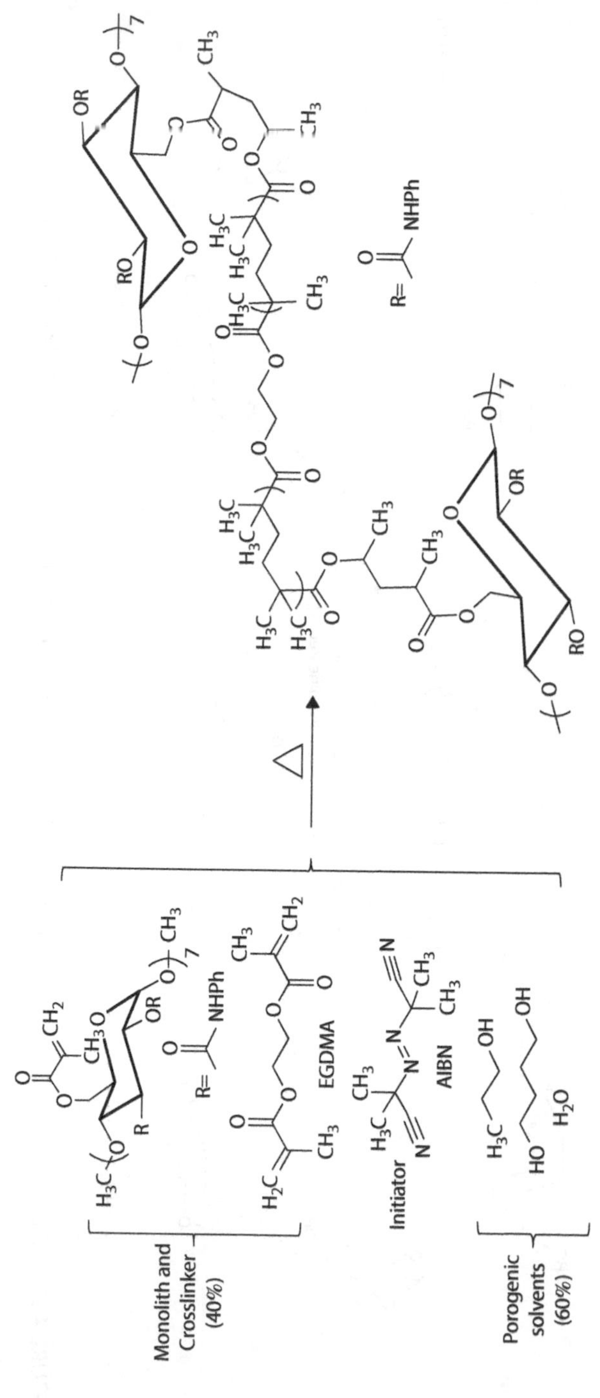

FIGURE 4.18 Schematic illustration for the preparation of β-CD-functionalized polymer monolith carried out inside capillaries.

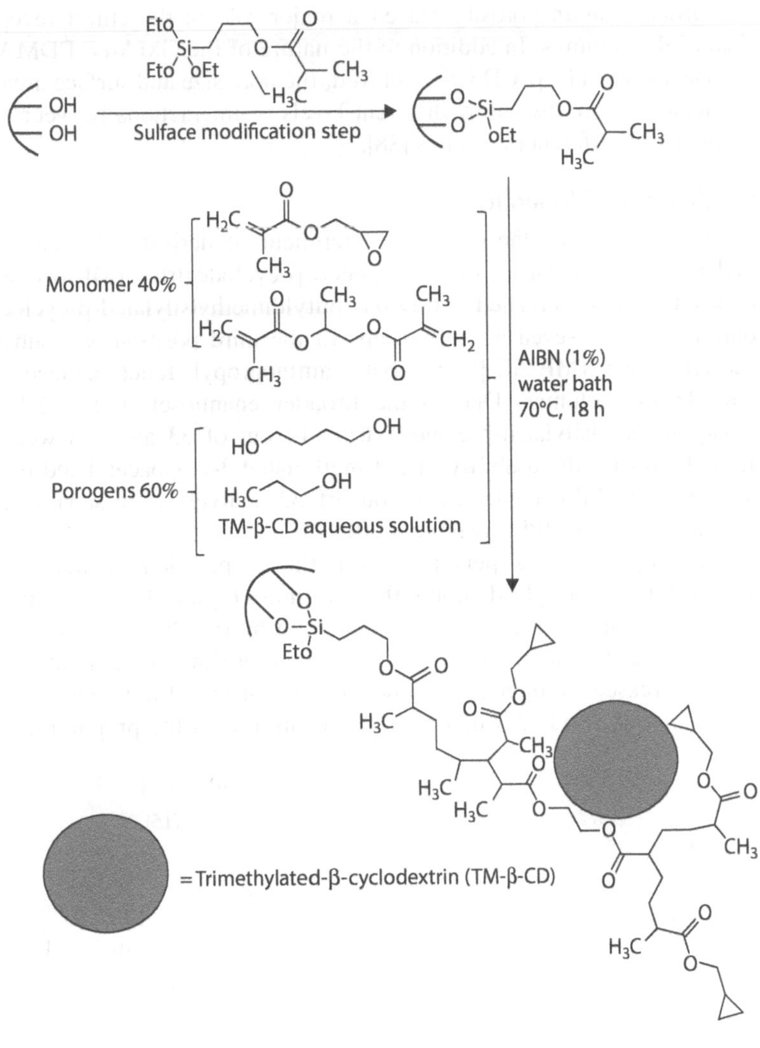

FIGURE 4.19 Schematic representation showing the *in situ* covalent attachment of the GMA-*co*-EDMA polymer matrix to the γ-MAPS anchored capillary inner wall while having TM-β-CD encapsulated within its network.

(Figure 4.19). The material characterization demonstrated that monolithic phases with higher concentration of TM-β-CD have relatively larger surface area, smaller pore size, and larger total pore volume compared to those with lower concentration TM-β-CD.

The new columns were efficient for the chiral separation of different classes of pharmaceuticals. The separation was done under reversed-phase conditions using mobile phase composed of methanol and water. Separation and resolution factors for the baseline resolved compounds were ranged from 1.07 to 1.42 and from 1.05 to 2.52, respectively. In contrast to the 2,3,6-*tris*(phenylcarbamoyl)-β-CD column described earlier, column porosity played a major role in the chiral recognition mechanism of the columns. In addition to the nature of the GMA-*co*-EDMA polymer backbone and the TM-β-CD CS employed, the pore size and surface area of the generated monolith can also allow different levels of interactions between the CS and enantiomers of different racemates [88].

4.3.3.2 Silica-Based Monoliths

Bayer et al. [89] reported the covalent attachment of derivatized cyclodextrins, 2,3-methylated 6-*O*-*tert*-butyldimethylsilylated-β-cyclodextrin (ME-β-CD) and 2,3-methylated-3-monoacetylated 6-*O*-*tert*-butyldimethylsilylated-β-cyclodextrin where only one of the seven methyl groups in the third position was substituted by an acetyl group (ME-AC-β-CD) onto aminopropyl functionalized monolithic silica HPLC columns. Due to the broader enantioselectivity, 2,3-methyl 6-*O*-*tert*-butyldimethylsilylated-β-cyclodextrin (14 out of 33 analytes were separated), it predominates the usability of 2,3-methylated-3-monoacetylated 6-*O*-*tert*-butyldimethylsilylated-β-cyclodextrin (7 out of 33 analytes were separated) as a chiral stationary phase in HPLC (Figure 4.20).

A *one-pot* approach was performed for the preparation of perphenylcarbamoylated β-CD-silica hybrid monolithic column (Figure 4.21). Capillary LC enantioseparations of 13 racemates were achieved by the Ph-β-CD-silica hybrid monolithic column. It demonstrated that the retention time and resolution of the enantiomers increased by increasing concentration of chiral monomer from 37.0 mg/mL, 45.0 mg/mL, and 53.3 mg/mL to 60.6 mg/mL in the prepolymerization

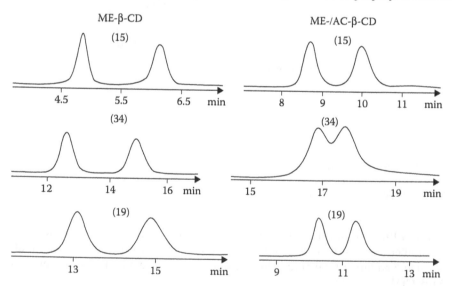

FIGURE 4.20 Comparison of the enantiomeric separation-ionone (15), 2,2,2-trifluoro-1-(9-anthryl)ethanol (34) and flavanone (19) on monolithic columns modified with 2,3- methylated 6-*O*-tert butyldimethylsilylated-cyclodextrin (ME-AC-CD) and with 2,3-methylated 6-*O*-tert butyldimethylsilylated-cyclodextrin (ME-CD) by HPLC in acetonitrile/0.1 M dihydrogen phosphate, pH 4. (From Bayer, M. et al., *J. Sep. Sci.*, 29, 1561–1570, 2006.)

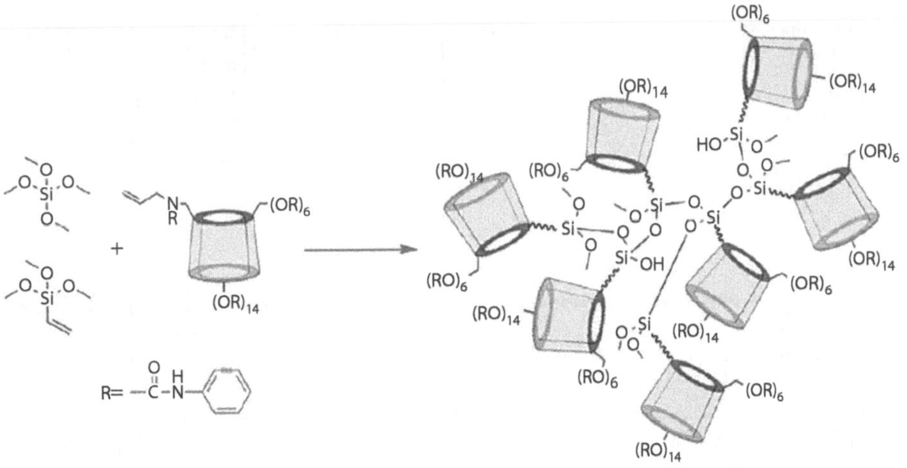

FIGURE 4.21 Synthetic approach of novel CD-CSPs. (From Zhang, Z.B. et al., *Anal. Chem.*, 83, 3616, 2011.)

mixture, and the baseline separation of compound 3 (Figure 4.22) could be achieved when the concentration of chiral monomer was 45.0 mg/mL. The limitation and difficulty of the use of water insoluble organic monomers have been overcome in the present study since the prepolymerization system consisted of organic solvent (methanol [MeOH] and *N,N*-dimethylformamide [DMF]). Therefore, various β-CD derivatives as well as other hydrophobic monomers could be involved in preparation of organic–silica hybrid monolithic columns with the *one-pot* process [90].

To explore the efficacy of silica-based monoliths versus polymer-based monoliths, Ghanem and coworkers prepared an immobilized β-cyclodextrin phenylcarbamate-based silica monolithic capillary columns using 2,3,6-*tris*(phenylcarbamoyl)-β-cyclodextrin-6-methacrylate as a functional monomer for the preparation of β-cyclodextrin(β-CD)-based silica via the sol–gel technique in fused silica capillary (Figure 4.23). Silica monolithic capillary functionalized with β-CD showed broader enantioselectivities. The polymer monolith is less time consuming and can be easily reproduced compared with the technically challenging silica monolith preparation [91].

4.3.4 Macrocyclic Antibiotic-Based Chiral Stationary Phases

4.3.4.1 Organic Polymer-Type Monoliths

A different backbone was prepared via the copolymerization of *N,N*-diallyltartardiamide, *N*-(hydroxymethyl)acrylamide, piperazine diacrylamide, and vinyl sulfonic acid for the enantioselective nano-liquid chromatographic separation of racemic pharmaceuticals. The prepared monolith bears diol groups on its surface, which is subject to oxidation to aldehyde groups via periodate treatment. The macrocyclic antibiotic, vancomycin, was immobilized on the activated monoliths by a reductive amination process (Figure 4.24) [92]. The prepared monolithic columns were used for the enantioselective separation of thalidomide, bupivacaine, and warfarin.

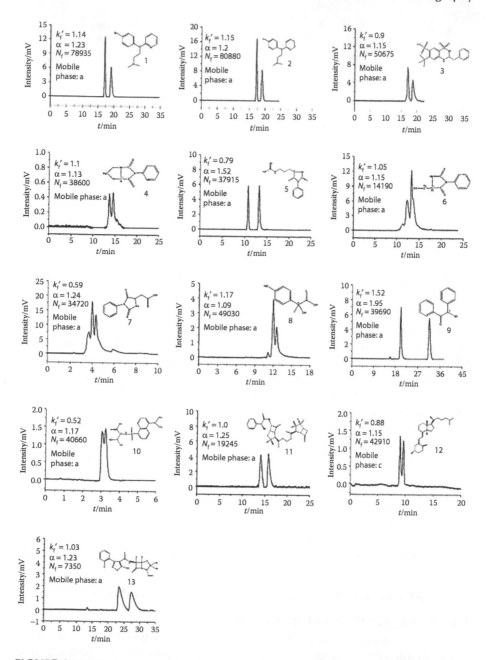

FIGURE 4.22 Enantioseparation of chiral compounds on the Ph-β-CD-silica hybrid mono-lithic column by cLC. Experimental conditions: effective length, 30 cm 75 μm i.d.; mobile phase, (a) MeOH/TEAA (pH = 4.2) = 60:40, MeOH/TEAA (pH = 4.2) = 80:20, Hexane/IPA = 90:10; flow rate, 120 nL/ min; detection wavelength, 214 nm. (From Zhang, Z.B. et al., *Anal. Chem.*, 83, 3616, 2011.)

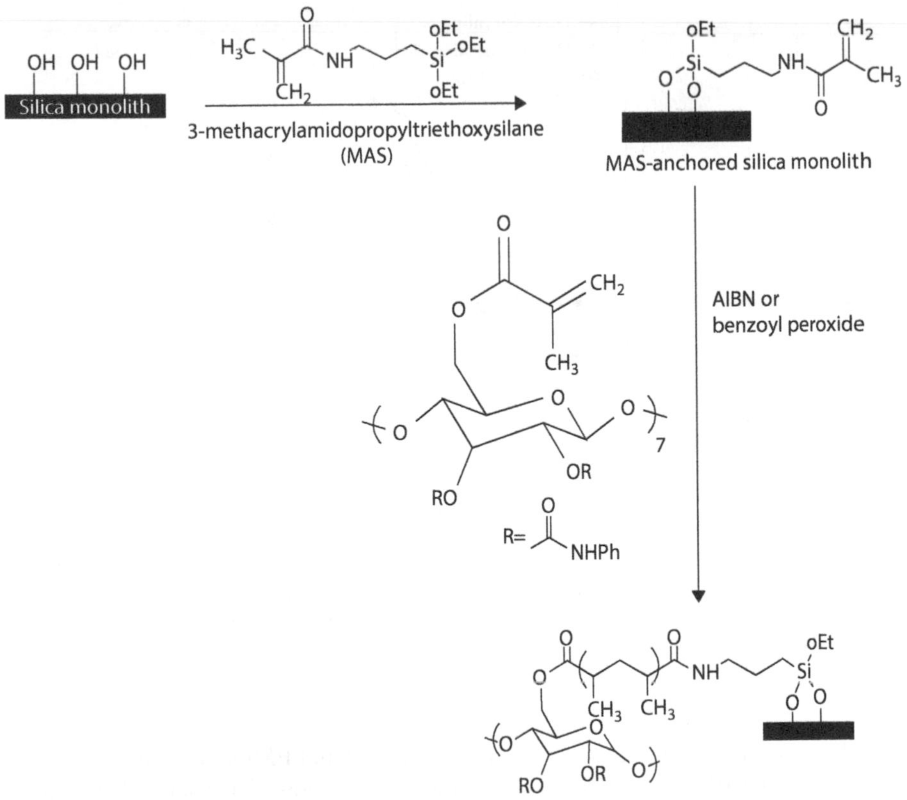

FIGURE 4.23 Schematic diagram for the preparation of β-cyclodextrin-modified silica monolith.

A RP-C18 monolithic column coated with *N*-(2-hydroxydodecyl)-vancomycin was prepared, successfully applied to the enantioseparation of dansylamino acids, and presented as a simple and inexpensive alternative for preparing CSPs [92].

The monolithic carbon nanotube (CNT) stationary phase was prepared by immobilization of pyrenyl derivative of neomycine A (PNA) using aqueous solution containing PNA at a concentration of 10 mM. The immobilized single-walled CNTs with a PNA and aminoglycoside antibiotic stationary phase were tested successfully for the enantioseparation of a series of 10 nonderivatized amino acids. Ultrafast separations in the range of seconds were achieved using high flow rates [93].

4.3.5 ALKALOID-BASED

4.3.5.1 Organic Polymer-Type Monoliths

A carbamoylated quinidine-based monolith, namely poly(*O*-9-[2-(methacryloyloxy)-ethylcarbamoyl]-10,11-dihydroquinidine-*co*-ethylenedimethacrylate(poly(MQD-*co*-EDMA)), was prepared and compared to poly(MQD-*co*-HEMA-*co*-EDMA)

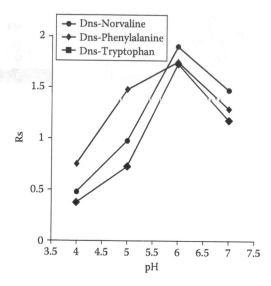

FIGURE 4.24 Effect of the pH on the resolution using dansyl-norvaline (•), dansyl-phenylalanine (♦), and dansyl-tryptophan (■) as model compounds. Stationary phase: RP18 monolithic column coated with *N*-(2-hydroxydodecyl)-vancomycin. Mobile phase: 0.1% TEAA, pH 4–7– methanol (50:50). Flow rate: 0.5 mL/min.

monolith for the enantioseparation of *N*-benzoylated and FMOC-derivatized amino acids. The poly(MQD-*co*-EDMA) monolithic column showed accelerated analytical process for the enantioseparation of NBD-derivatized amino acids when compared to quinine or quinidine-based packed columns. The impact of comonomer HEMA has been studied where no significant influence on the enantiorecognition ability of the quinidine-based monolith was reported [94].

To enhance column permeability, efficiency and selectivity for micro-HPLC, a capillary monolithic column containing *O*-9-[2-(methacryloyloxy)-ethylcarbamoyl]-10,11-dihydroquinidine(MQD) as CS was prepared. The monolithic column was used to successfully enantioseparate a wide range of *N*-derivatized amino acids including alanine, methionine, leucine, threonine, phenylalanine, serine, valine, isoleucine, tryptophan, and cysteine. The optimized poly(MQD-*co*-HEMA-*co*-EDMA) monolithic column showed excellent morphology, reproducibility, good permeability, mechanical and chemical stability, and satisfactory chromatographic performance in micro-HPLC. The influence of organic solvents composition, the concentration of buffer, and the apparent pH of the mobile phase on the retention and enantioseparation of *N*-derivatized amino acids seem to confirm that both hydrophobic and electrostatic interactions are responsible for the retention of these acidic analytes. The optimized monolithic column was finally applied to the enantioseparation of a wide range of *N*-derivatized amino acids with high selectivity and good resolution, especially for 3,5-dinitrobenzoyl-amino acids (DNB-amino acids) and 3,5-dichlorobenzoyl-amino acids (DCIB-amino acids) (Figure 4.25) [95].

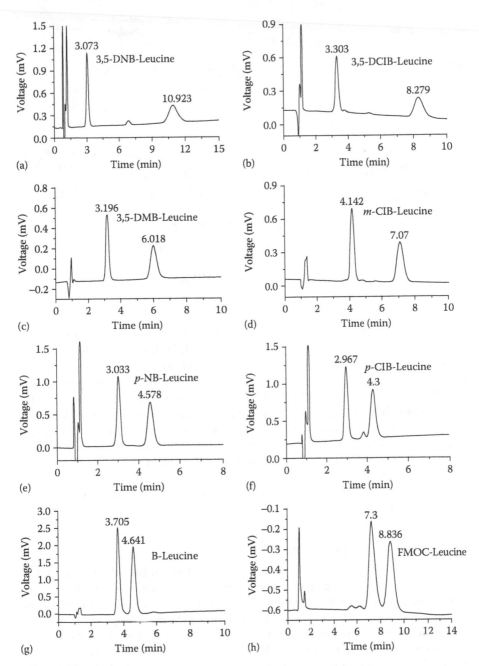

FIGURE 4.25 Enantioseparation of *N*-derivatized leucine derivatives on the poly(MQD-*co*-HEMA-*co*-EDMA) monolithic column. Conditions: column dimensions: 150 mm. 100 μm I.D.; mobile phase: (i) ACN/0.1 M ammonium acetate (80/20, v/v) (apparent pH = 5.3) for all analytes (a–g) except FMOC-Leucine, (ii) ACN/0.1 M ammonium acetate (50/50, v/v)-(apparent pH = 5.3) for FMOC-Leucine (h); UV detection wavelength: 254 nm; flow rate: 1 μL/min; injection volume: 20 nL.

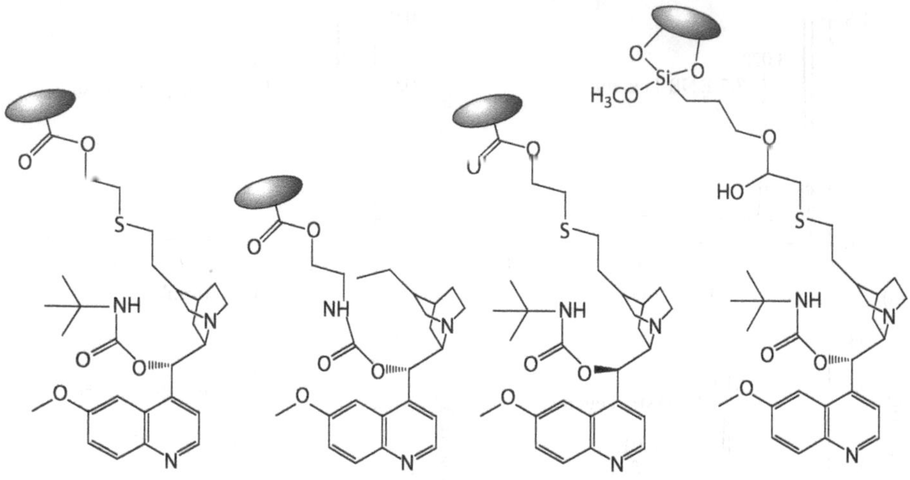

FIGURE 4.26 Surface chemistry of four prepared MBQD monolithic columns.

In a recent study, a novel functionalized monolith was synthesized using the O-9-(tert-butylcarbamoyl)-11-(2-hydroxyethylthio)-10,11-dihydroquinidine as CS poly(MBQD-co-HEMA-co-EDMA) and compared to the previously prepared poly(MQD-co-HEMA-co-EDMA) monolithic column. Surface chemistry of the four monolithic columns was described in Figure 4.26. The monolithic column exhibited excellent morphology, high mechanical stability, good permeability, reproducibility, and satisfactory separation efficiency of 47 N-derivatized amino acids in micro-LC (Figure 4.27). The comparison between the poly(MBQD-co-HEMA-co-EDMA) and poly(MQD-co-HEMA-co-EDMA) monoliths revealed higher enantioselectivity and diastereoselectivity for the new MBQD monomer with a more bulky tert-butyl carbamate residue on quinidine against MQD monomer bearing a linear alkyl chain at the carbamate group [96].

4.3.5.2   Silica-Based Monoliths

There are two fundamental methods to modify monolithic silica gels used in separation science. In the first method, particulate silica materials are chemically modified using batch reactions followed by the packing of the modified materials into the column. For such a reaction scheme, the particles are dispersed in a solvent and after addition of the silane reaction partner, the suspension is refluxed under inert atmosphere for several hours. The second method involved the *in situ* modification of silica gel already packed into the stainless steel tubing. This is performed by pumping the reaction partner solution at elevated temperature through the column. The *t*-BuCQN stationary phase was prepared by the reaction of (8S,9R)-O-9-(tert-butylcarbamoyl)-quinine with mercaptopropyl-modified silica monolith column using a radical addition procedure (Figure 4.28) [97].

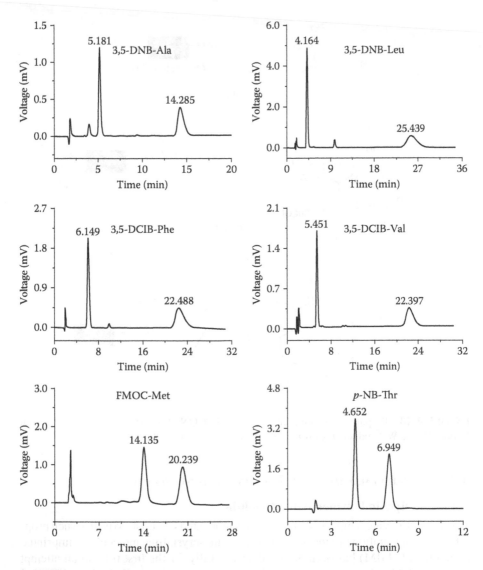

FIGURE 4.27   Enantioseparation of *N*-derivatized amino acids on the poly(MBQD-*co*-HEMA-*co*-EDMA) monolithic column. Micro-LC conditions: 0.1 M ammonium acetate/ACN (20/80 v/v, pH 5.3) mobile phase for all analytes except FMOC-Met; 0.1 M ammonium acetate/ACN (40/60 v/v, pH 5.3) for FMOC-Met, 800 nL/min flow rate, 254 nm.

Based on the selector–selectand model of the *tert*-butyl-carbamoylquinine modification using *N*-(3,5-dinitrobenzoyl)leucine as a model enantiomer (Figure 4.29), there are at least four different possible interactions, namely ionic, hydrogen bond, π–π, and steric interaction via van der Waals increments. The *tert*-butylcarbamoylquinine chiral anion-exchanger covalently bonded to monolithic silica support was used as a stationary phase for the enantioseparation of several *N*-derivatized amino acids and suprofen [97].

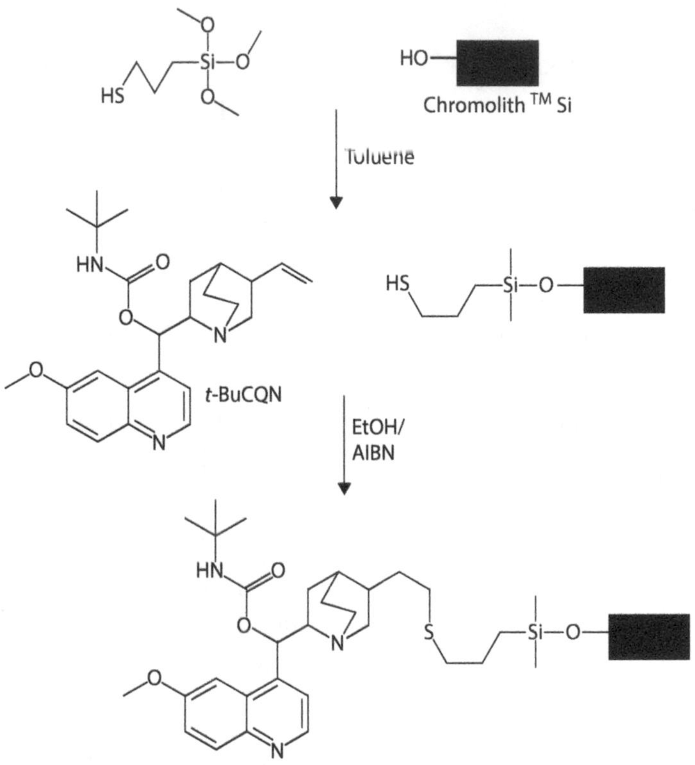

FIGURE 4.28 Preparation of monolithic silica *t*-BuCQN chiral stationary phases. (From Lubda, D. and W. Lindner, *J. Chromatogr. A.*, 1036, 135–143, 2004.)

4.3.6 MOLECULARLY IMPRINTED POLYMER-BASED MONOLITHIC COLUMNS

4.3.6.1 Organic Polymer-Type Monoliths

The ability to create template-shaped cavities motivated Chemists to utilize such property in the preparation of monoliths. Thus, a liquid-crystalline molecularly imprinted monolith (LC-MIM) has been prepared successfully for the first time in an attempt to develop highly specific chiral stationary phase for cinchona alkaloids. Polymethyl methacrylate (PMMA) or polystyrene (PS) as a crowding agent, 4-cyanophenyl dicyclohexyl propylene (CPCE) as liquid crystal monomer, and hydroquinidine as pseudotemplate were used. A low level of cross-linker (26%) was sufficient to achieve molecular recognition on the crowding-assisted LC-MIM due to the physical cross-linking of mesogenic groups in place of chemical cross-linking. PMMA displayed better crowding effect. Acetonitrile and the pH of the mobile phase were detrimental for the chiral recognition of the liquid-crystal-based monolith. LC-MIM was proved as a potential technique to prepare specific and high-performing CSPs [98].

In a comparison between imprinted and nonimprinted molecularly monolithic columns, a monolithic molecularly imprinted polymer (monolithic MIP) using dibenzoyl-D-tartaric acid (D-DBTA) as template, acrylamide (AM) monomer,

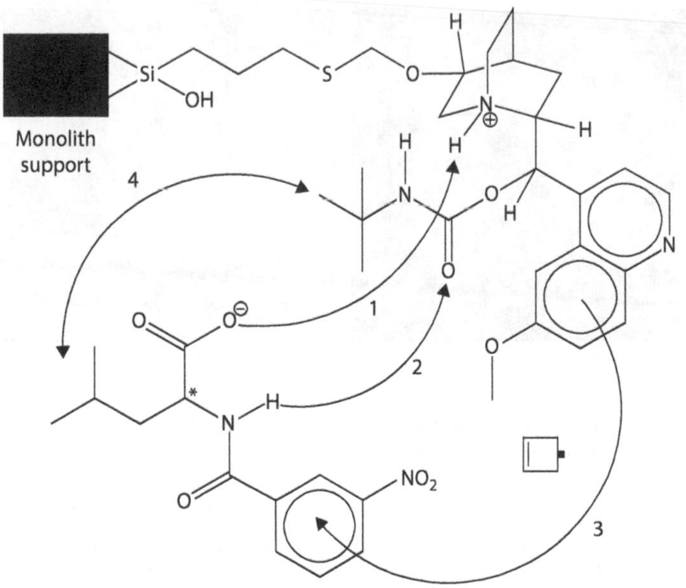

FIGURE 4.29 Selector–selectand interaction model of *tert*-butyl-carbamoylquinine (1*S*, 3*R*, 4*S*, 8*S*, 9*R*) modification and *N*-(3,5-dinitrobenzoyl)leucine. (1) ionic interaction, (2) hydrogen bond, (3) π–π interaction, and (4) steric interaction. (From Lubda, D. and W. Lindner, *J. Chromatogr. A.*, 1036, 135–143, 2004.)

free-radical initiator (AIBN), and cross-linker (EGDMA) was prepared within the confines of a stainless-steel chromatographic column. Nonimprinted molecularly monolithic columns (NIP) for the comparison experiment was prepared without the addition of the template. The chiral separation of DBTA enantiomers was successfully achieved on the molecularly imprinted polymer (MIP), whereas no enantioseparation was noticed on the monolithic nonimprinted polymer (NIP). In this study, AM was proved to be the better monomer compared to MMA. Furthermore, the best performance was achieved via template:monomer (1:6), monomer:cross-linker (1:4), and toluene:dodecanol (1:8) [99].

In a trial to understand the difference between low- and high-density MIP, (*S*)-naproxen-imprinted monolith was prepared in both low-density (30% [w/w] monomers) and high-density (40% [w/w] monomers) polymers (Figure 4.30). The resultant low-density MIPs monolithic column showed shorter analysis time (<10 min) and higher column efficiency due to improved mass transfer and good column permeability. In this study, the ratio of the weight fractions of monomers compared to that of the solvent significantly affected column efficiency of MIP monoliths [100].

In a further study, the use of both ionic liquid and the pivot strategy for chiral separation by molecular imprinting was reported. (*R*)-mandelic acid was used as a template in a polymerization mixture composed of 4-vinylpyridine and EDMA along with a ternary porogenic system composed of the ionic liquid 1-butyl-3-methylimidazolium tetrafluoroborate, DMSO, and DMF in the presence of different metal ions such as cobalt, copper, nickel, and zinc. The presence of ionic liquid in the porogen

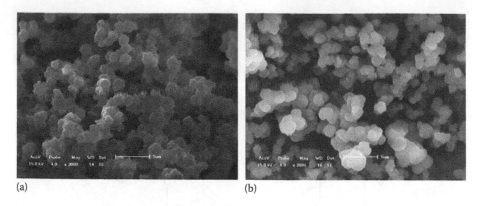

FIGURE 4.30 Scanning electron micrographs of the high-density (P4) (a) and low-density (P7) (b) monolithic imprinted polymers.

increased column efficiency and selectivity of metal ion-mediated MIP monoliths. Using metal ions as pivot is considered as an effective method for preparing a chiral stationary phase based on MIP. When used between the template and functional monomer (metal-mediated self-assembly) (Figure 4.31), it offers stronger coordination bond replacing the weak interaction (for example hydrogen bonding) between the monomer and template. This is of particular importance for self-organization between templates and functional monomers in a polar environment. The template consisting of 63:6:31 [BMIM] BF4:DMF:DMSO showed the optimum performance. The addition of metal ions or template to the prepolymerization mixture provoked smaller throughpores. Co^{2+} was a better pivot than other metal ion [101].

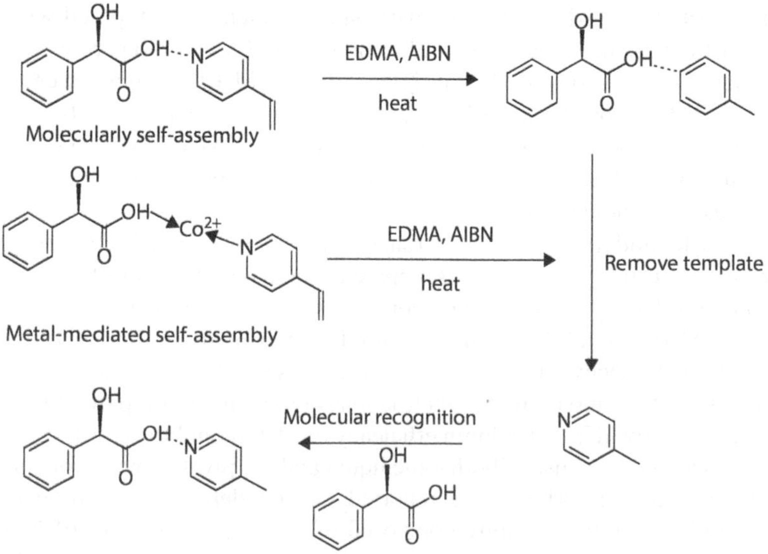

FIGURE 4.31 Schematic representation of the preparation of an MIP by use of metal ions as pivot, and of molecular recognition by the MIP.

4.3.6.2 Silica-Based Monoliths

The MIP has also been utilized in silica monoliths. Thus, in 2007, Ou et al. developed a silica monolith derivatized by molecularly imprinted polymer. L-tetrahydropalmatine (*l*-THP) and (5*S*,11*S*)-(-)-Tröger's base (*S*-TB) were used as templates, then the derivatized monolith (γ-methacryloxypropyl triemethoxysilane-modified monolith) was anchored to the silica via copolymerization with vinyl groups. The preparation conditions such as monomer concentration, temperature, and time of polymerization were studied, and the separation of tetrahydropalmatine and tröger's base using CLC was successful. The results were compared to the separations achieved by a similar organic monolith where silica-based monolith was superior in terms of column efficiency and stability [102].

Propranolol imprinted beaded, silica-grafted, and grounded MIP have been prepared and used in the separation of nine β-blocker using HPLC and TFC. Beaded and grounded MIP materials displayed a degree of chiral separation for all tested β-blockers. The beaded MIP was better in relation to flow properties and peak shape, whereas silica-grafted MIP showed some separations in five of the drugs and a large improvement in peak shape and analysis times compared with both ground and beaded MIPs [103].

4.3.7 MISCELLANEOUS

4.3.7.1 Organic Polymer-Type Monoliths

In 2014, Chiroli et al. have reported the first example of a chiral organocatalyst immobilized onto a monolithic polymer by radical copolymerization of divinylbenzene and a properly modified enantiomerically pure imidazolidinone inside a stainless steel column in the presence of dodecanol and toluene as porogens. Such polymer has been utilized for organocatalyzed cycloadditions between cyclopentadiene and cinnamic aldehyde under continuous-flow conditions in both catalytic reactor and chiral column. Despite no separation of enantiomers has been reported using the prepared chiral column, it represented an innovative technique for chiral column preparation, which could later utilized in chiral analysis (Figure 4.32) [104].

4.3.7.2 Silica-Based Monoliths

Silica monoliths were functionalized with anion- or cation-exchange-type CS via immobilization of *trans*-(1*S*, 2*S*)-2-(*N*-4-allyloxy-3,5-dichlorobenzoyl)amino cyclohexane sulfonic acid. The separation of enantiomers of various basic pharmaceuticals such as β-blockers and β-sympathomimetics (Figure 4.33) was investigated in both aqueous and nonaqueous mobile phase. Aqueous conditions in nano-HPLC showed the best results in terms of selectivity and resolutions [105].

In 2009, a triazole-linked brush-type chiral stationary phase was prepared by Slater et al. [106] using copper-catalyzed azide-alkyne cycloaddition click chemistry. Thus, the silica surface was first modified with 3-(azidopropyl)trimethoxysilane where the azide functionalities were located. This was followed by the immobilization of the alkyne containing proline derived CS (*N*-(*N*-(5-hexynoyl)prolinoyl)-5-aminoindan) in the presence of a copper (I) iodide. Both silica monolith and porous silica beads were prepared and investigated for the separation of four *N*-(3,5-dinitrobenzoyl) amino acid dialkyl amide where excellent enantioselectivity toward π-acidic amino acid amide derivatives was observed with both monoliths and beads. In contrary to

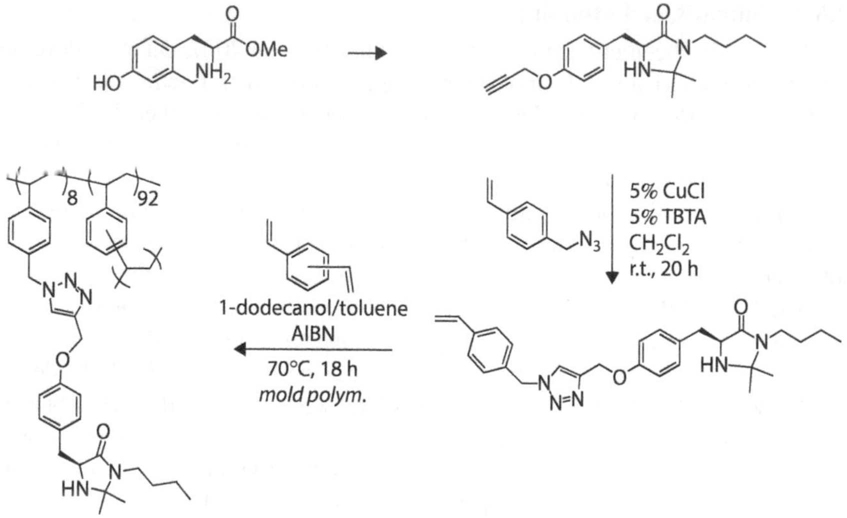

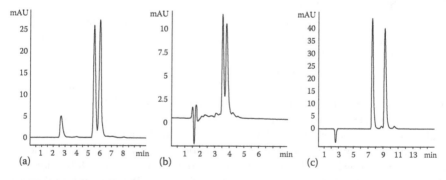

FIGURE 4.32 Preparation of the monolithic reactor containing the chiral imidazolidinone catalyst.

the residual azide functionality, which does not contribute to nonspecific retention, the unreacted silanol groups lead to extensive nonspecific interactions and poor behavior.

In a further study, a low-molecular mass CS namely (R)-acryloyloxy-β-β-dimethyl-γ-butyrolactone was bonded to a modified silica-based monolith to form a new brush-type chiral stationary phase for micro-HPLC separation. The monolithic silica surface was easily modified with an anchor group containing an amide moiety followed by the copolymerization with lactone acrylate derivative to afford a chiral α-lactone-functionalized silica monolith. The chiral monolith was examined for the separation of a set of compounds where baseline separation was obtained for a set of secondary alcohols, whereas partial separation was achieved for 1-phenylethylamine [107] (Table 4.2).

FIGURE 4.33 Enantiomeric separation of pronethalol (a), salbutamol (b), and mefloquine-*t*-butylcarbamate (c) by nano-HPLC () with a SCX-modified silica monolith. Mobile phase: acetonitrile/methanol (80:20, v/v) containing 25 mM formic acid and 12.5 mM (R,S)-2-amino-1-butanol). Conditions nano-HPLC: capillary 33.5 cm. 100 m I.D.; flow rate: 0.35 μL/min. (From Preinerstorfer, B. et al., *Electrophoresis*, 29, 1626–1637, 2008.)

TABLE 4.2

Chiral Selectors Immobilized on Monolithic LC Stationary Phases

Chiral selector Class	Monolithic Type	Chiral Selector	Mobile Phase	Analytes	LC Type	Ref.
Protein and peptides	Organic polymer	Human serum albumin (HSA)	Pure water	d,l-amino acids (AAs) (d,l-Trp, d,l-Phe, and d,l-Tyr)	HPAC	[55]
		Human serum albumin (HSA)	0.067 M potassium phosphate buffer containing 0.5% isopropanol	(R/S)-warfarin and D/L-tryptophan	HPAC	[43,56]
		Human serum albumin (HSA)	20 mM sodium phosphate, pH 7.2, methanol/water (0.1% TFA) 30:70 v/v	D/L-tryptophan and dl-kynurenine	CLC	[57]
		Candida antarctica lipase B (CALB)		Celiprolol, carbuterol, atenolol, normetanephrine, nomifensine, alprenolol, cizolertine, carbinol, bromoglutethimide, desmethylcizolertine, chlorpheneramine, 4-hydroxy-3-methoxymandelic acid, sulconazole, and miconazole	Nano-HPAC	[20]
	Silica	Human serum albumin (HSA)	0.067 M potassium phosphate buffer containing 0.5% isopropanol and 5 mM octanoic acid	D/L-tryptophan or R/S-warfarin and R/S-ibuprofen	HPAC	[61]

(Continued)

TABLE 4.2 (Continued)

Chiral Selectors Immobilized on Monolithic LC Stationary Phases

Chiral selector Class	Monolithic Type	Chiral Selector	Mobile Phase	Analytes	LC Type	Ref.
		Human serum albumin (HSA)	0.067 M potassium phosphate buffer	D/L-tryptophan, R/S-warfarin and R/S-ibuprofen	HPAC	[62]
		Trypsin	100 mM phosphate buffer (pH 7.0)	Somatostatin and myoglobin	HPLC–UV– ESI–MS–MS	[63]
		α1-acid glycoprotein AGP	0.067 M potassium phosphate buffer (pH 7.4)	R/S-propranolol, epinephrine, isoproterenol, phenylbutazone, carbamazepine, pindolol, lidocaine, perphenazine, quinidine, imipramine, bupivacaine, and trifluoperazine	HPLC	[64]
		α1-acid glycoprotein AGP	(0.067 M KPB, 3% isopropanol) (0.067 M KPB 5% isopropanol)	R/S-propranolol and R/S-warfarin	HPLC	[66]
		Cyclic hexapeptide molecule	Phosphate buffer (pH 5.60)	Dan-ala, dan-val, dan-nor, dan-leu, dan-phe, dan-try, etodolac flobulin, flurbiprofen, ibuprofen and naproxen, and sulpiride	HPLC	[67]
		Polyproline derivatives	Heptan:isopropano l90:10	22 Racemates	HPLC	[68]
		Octaproline	Heptan:isopropano l90:10	29 racemates	HPLC	[69]

(Continued)

TABLE 4.2 (Continued)
Chiral Selectors Immobilized on Monolithic LC Stationary Phases

Chiral selector Class	Monolithic Type	Chiral Selector	Mobile Phase	Analytes	LC Type	Ref.
		Phenylalaninamide, prolinamide, alaninamide	acetonitrile–[50 mM NH Ac–0.50 mM 4 Cu(Ac)] (7:3), acetonitrile–[0.1 M NH Ac–0.25 mM Cu(Ac)] 2 4 2(7:3)	Dns-DL-amino acids and some hydroxy acids	μ LC, μ-HPLC	[70,71]
		N-decyl-L-4-hydroxyproline, N-hexadecyl-L-4-hydroxyproline, or N-2-hydroxydodecyl-L-4-hydroxyproline	0.1 mM Cu(II)sulfate	Gly-Asp, Gly-Leu, Gly-Met, Gly-Nle, Gly-Nva, Gly-Phe, Gly-Thr, Gly-Trp, Gly-Val	HPLC	[72,73]
		Penicillin G acylase	50 mM phosphate buffer (pH 7.0)	14 racemates	HPLC	[74]
		Penicillin G acylase	50 mM phosphate buffer (pH 7.0)	(S)-Ketoprofen, trometamol	CLC	[75]
Polysaccharides	Silica	Cellulose tris(3,5-dimethyl-phenylcarbamate) (CDMPC)	n-hexane/2-propanol, 90/10 (v/v)	2,2,2-trifluoro-1-(9-anthryl) ethanol, benzoin, 2,29-dihydroxy-6,69-dimethylbiphenyl, trans-cyclopropanedicarboxylic acid dianilide, flavanone, trans-stilbene oxide	CLC	[76]

(Continued)

TABLE 4.2 (Continued)
Chiral Selectors Immobilized on Monolithic LC Stationary Phases

Chiral selector Class	Monolithic Type	Chiral Selector	Mobile Phase	Analytes	LC Type	Ref.
		Cellulose 2,3-bis(3,5-dimethylphenylcarbamate)-6-(3,5-dimethylphenylcarbamate)(2-methacryloyloxyethylcarbamate), cellulose 2,3-bis(3,5-dichlorophenylcarbamate)-6-(3,5-dichlorophenylcarbamate)(2-methacryloyloxyethylcarbamate), and amylose 2,3-bis(3,5-dimethylphenylcarbamate)-6-(3,5-dimethylphenylcarbamate)(2-methacryloyloxyethylcarbamate)	n-Hexane/2-propanol in the ratio 90:10 v/v	Tröger's base, trans-stilbene oxide, benzoin, 1,2,2,2-tetraphenylethanol, 2-phenylcyclohexanone, 2,2,2-trifluoro-1-(9-anthryl)ethanol, cobalt(III) tris(acetylacetonate), flavanone, trans-cyclopropandicarboxylic acid dianilide and 2,29-dihydroxy-6,69-dimethylbiphenyl	CLC	[77]
		Amylose tris-(3-chlorophenylcarbamate)	n-hexane/2-propanol 90:10 v/v	1-Acenaphthenol, Carprofen, celiprolol, cizolirtine, carbinol, miconazole, tebuconazole, 4-hydroxy-3-methoxymandelic acid,1-indanol, 1-(2-chlorophenyl)ethanol, 1-phenyl-2-propanol, flavanone, 6-hydroxyflavanone, 4-bromogluthethimide, pentobarbital, aminoglutethimide, naftopidil and thalidomide	HPLC, nanc-HPLC	[78]

(Continued)

TABLE 4.2 (Continued)
Chiral Selectors Immobilized on Monolithic LC Stationary Phases

Chiral selector Class	Monolithic Type	Chiral Selector	Mobile Phase	Analytes	LC Type	Ref.
		Amylose 3,5-dimethylphenylcarbamate (ADMPC) and cellulose 3,5-dichlorophenylcarbamate (CDCPC)	Methanol, ethanol, 2-propanol and acetonitrile modified with 0.1% formic acid	Flavanone and its chiral derivatives	HPLC	[79]
	Hybrid	Cellulose *tris* (3,5-dimethylphenylcarbamate) (CDMPC)	*n*-hexane/2-propanol (90/10, v/v)	Benzoin, Alprenolol, *trans*-stilbene oxide, praziquantel, and flavanone	HPLC	[80]
Oligosaccharides	Organic polymer	Hydroxypropyl-β-cyclodextrin (HP-β-CD)	20/80 methanol/water and 20/80 acetonitrile/water, both buffered with 0.1% triethylamine-acetate	Nomifensine, praziquantel, 5-methyl-5-phenylhydantoin and alprenolol	nano-LC	[82]
		Mono-(6-azido-6-deoxy)-β-cyclodextrin	Methanol and borate buffer	Flavanone enantiomers	CLC	[83]
		β-cyclodextrin	(0.3% TEAA, pH 4.9)/ methanol (85/15, v/v), MeOH/(0.3% TEAA, pH 5.4), acetonitril/ (0.3% TEAA, pH 6.8)	Chlorthalidone, propranolol, aminoglutethimide, benzoin, amlodipine, chlorpheniramine, D,L-phenylalanine, ibuprofen	HPLC	[84]

(Continued)

TABLE 4.2 (*Continued*)
Chiral Selectors Immobilized on Monolithic LC Stationary Phases

Chiral selector Class	Monolithic Type	Chiral Selector	Mobile Phase	Analytes	LC Type	Ref.
		Mono-6-amino-6-deoxy-β-CD,mono-6-ethylenediamine-6-deoxy-β-CD, mono-6-hexamethylenediamine-6-deoxy-β-CD	Acetonitrile/20mMKH$_2$PO$_4$ (pH 4.2)(70/30, v/v)	Suprofen, idoprofen, ibuprofen 2-(3-chlorophenoxy)-propionic acid, phenoxypropionic acid, phenylpropionic acid	HPLC	[85]
		Ethylenediamine-β-cyclodextrin	MeOH/0.5% (v/v) TEA (30:70, v/v),	Ibuprofen racemate	HP_C	[86]
		2,3,6-*tris*(Phenylcarbamoyl)-cyclodextrin-6-methacrylate	0.1% TFA in water (v/v) and methanol (v/v)	β-blockers, α-blockers, anti-inflammatory drugs, antifungal drugs, dopamine antagonists, dopamine reuptake inhibitor, catecholamines, sedative hypnotics, diuretics, antihistaminics, anticancer drugs, antiarrhythmic drugs, and miscellaneous drugs	Nano-HPLC	[87]
		2,3,6-*tris*(Phenylcarbamoyl)-CD	0.1 %TFA in water (v/v) and methanol (v/v) or 0.1% TFA in water (v/v) and acetonitrile	β-blockers, α-Blockers, anti-inflammatory drugs, antifungal drugs, dopamine antagonists, dopamine reuptake inhibitor, catecholamines, sedative hypnotics, diuretics, antihistaminics, anticancer drugs, antiarrhythmic drugs, and miscellaneous drugs	Nanc-HPLC	[88]

(Continued)

(Continued)

TABLE 4.2 (*Continued*)
Chiral Selectors Immobilized on Monolithic LC Stationary Phases

Chiral selector Class	Monolithic Type	Chiral Selector	Mobile Phase	Analytes	LC Type	Ref.
	Silica	2,3-Di-*O*-methyl-3*-OH-69-*O*-(oct-7-enyl)-β-CD, 2,3-Di-*O*-methyl-3*-OH-6-*O*-TBDMS-69-*O*-(oct-7-enyl)-β-CD, 2,3-Di-*O*-methyl-3*-*O*-acetyl-6-*O*-TBDMS-69-*O*-(oct-7-enyl)-β-CD, and 2,3-Di-*O*-methyl-3*-*O*-acetyl-6-*O*-TBDMS-69-*O*-(7-epoxyoctyl)-β-CD	ACN/potassium dihydrogen phosphate buffer, 0.1 M, pH 4.0	α-Ionone, carvone, piperitone, *trans*-a-irone, *cis*-a-irone, flavanone, 2-hydroxyflavanone, 4-hydroxyflavanone, 6-hydroxyflavanone, benzoin, thalidomide, secobarbital, pentobarbital, atenolol, propranolol, mandelic acid, ibuprofen, mecoprop, *trans*-stilbene oxide, 1,1-Bi(2-naphthol), 2,2,2-trifluoro-1-(9-anthryl) ethanol, oxazepam, lorazepam, 1-phenylethanol, 1-phenyl-1-propanol, 1-phenyl-2 propanol, 4-phenylcyclohexanone, oxybutynin, warfarin, pemoline, 1-(1-naphthyl) ethanol, 9 hydroxyoctadecadienoic acid, and 5-hydroxyeicosatetraenoic acid	HPLC	[89]

TABLE 4.2 (Continued)

Chiral Selectors Immobilized on Monolithic LC Stationary Phases

Chiral selector Class	Monolithic Type	Chiral Selector	Mobile Phase	Analytes	LC Type	Ref.
		Perphenylcarbamoylated β-CD-	MeOH/TEAA (pH 4.2) 60:40, MeOH/TEAA (pH 4.2) 80:20, Hexane/IPA = 90:10	13 racemates	(cLC)	[90]
		2,3,6-tris(phenylcarbamoyl)-cyclodextrin-6-methacrylate	0.1% TFA in water (v/v) and methanol (v/v)	β-Blockers, α-blockers, Anti-inflammatory drugs, antifungal drugs, dopamine antagonists, dopamine reuptake inhibitor, catecholamines, sedative hypnotics, diuretics, antihistaminics, anticancer drugs, antiarrhythmic drugs, and miscellaneous drugs	Nano-HPLC	[91]
Macrocyclic antibiotics	Organic polymer	N-(2-hydroxydodecyl)-vancomycin	0.1% TEAA, pH 6–methanol (50:50)	11 Dansyl-amino acids	HPLC	[92]
		Pyrenyl neomycine A (PNA)	H_2O/$CuSO_4$, 1 mM	Tryp-Tyr-Phe-Glu-Asp-Gln-Asn-Val-Ser-Ala-	HPLC	[93]
Alkaloid	Organic polymer	O-9-[2-(methacryloyloxy)-ethylcarbamoyl]-10,11-dihydroquinidine	ACN/0.1 M ammonium acetate (80/20; v/v)	N-Protected amino acids, including 3,5-DNB,3,5-DCIB, FMOC, 3,5-DMB, p-NB, m-CIB, and p-CIB derivatives	micro-LC	[94]

(Continued)

TABLE 4.2 (Continued)
Chiral Selectors Immobilized on Monolithic LC Stationary Phases

Chiral selector Class	Monolithic Type	Chiral Selector	Mobile Phase	Analytes	LC Type	Ref.
		O-9-[2-(methacryloyloxy)-ethylcarbamoyl]-10,11-dihydroquinidine (MQD)	ACN/0.1 M ammonium acetate (80/20; v/v)	N-derivatized amino acids including alanine, leucine, methionine, threonine, phenylalanine, valine, serine, isoleucine, tryptophan, and cysteine	micro-HPLC	[95]
		O-9-(tert-butylcarbamoyl)-11-[(2-methacryloyloxy)ethylthio]-10,11-dihydroquinidine (MBQD)	20% 0.1 M ammonium acetate and 80% ACN	47 N-derivatized amino acids	μ LC	[96]
	Silica	O-9-(tert-Butylcarbamoyl) quinidine (t-BuCQD)	ACN/0.1 M ammonium acetate (80/20; v/v)	N-derivatized amino acids ???	micro-LC	[97]
	Organic polymer	Hydroquinidine	Acetonitrile/50 mmol L–1 acetate (pH 5.0) (70:30, v/v)	Quinidine and quinine	HPLC	[98]
MIP		Dibenzoyl-D-tartaric acid (D-DBTA)	Acetonitrile (100%), acetonitrile–water (90:5, v/v) and acetonitrile–water (90:10, v/v)	rac-DBTA and rac-DTTA	LC	[99]
		(S)-Naproxen	Acetonitrile–50 mM NaAc/HAc buffer (pH = 3.6), 90/10 (v/v)	Racemic naproxen	LC	[100]
		rac-Mandelic acid	–NaAc–HAc buffer (50 mmol L–1, pH 3.6) (90:10, v/v)	rac-mandelic acid	HPLC	[102]

(Continued)

TABLE 4.2 (Continued)

Chiral Selectors Immobilized on Monolithic LC Stationary Phases

Chiral selector Class	Monolithic Type	Chiral Selector	Mobile Phase	Analytes	LC Type	Ref.
		Propranolol	Acetonitrile:water (70:30) with 0.01% TFA	Acebutolol, alprenolol, atenolol, carvedilol, metoprolol, nadolol, oxprenolol, pindolol, and propranolol	HPLC	[103]
Miscellaneous	Organic polymer	Imidazolidinone(organocatalyst)	Acetonitrile:water (95:5)	cyclopentadiene and cinnamic aldehyde	HPLC	[104]
	Silica	trans-(1S,2S)-2-(N-4-Allyloxy-3,5-dichlorobenzoyl) aminocyclohexanesulfonic acid	(a) ACN-MeOH, (b) ACN-H2O, (c) MeOH-ACN, and (d) MeOH-H2O, always 80:20 (v/v)	Mefloquine, mefloquine-t-butylcarbamate, clenbuterol, pronethalol, salbutamol, talinolol, sotalol and rimiterol	Nano-HPLC	[105]
		Alkyne-containing chiral selector (N-(N-(5-hexynoyl) prolinoyl)-5-aminoindan)	10% 2-propanol in hexanes	p-acidic amino acid amide derivatives	LC	[106]
		(R)-acryloyloxy-β-β-dimethyl-γ-butyrolactone	n-hexane/2-propanol (98:2 v/v)	1-indanol, 1-Phenylethylamine, α-phenyl glycinol, 1-(4-Bromophenyl) ethanol, and 1-(2-chlorophenyl) ethanol	μHPLC	[107]

4.4   CONCLUSION

This chapter summarized the recent development of CSs' immobilization on HPLC monolithic stationary phases. Special emphasis was given to the description of different types of techniques used for immobilization of CSs on monolithic matrices. The technologies used along with the structural and chemical properties of monoliths demonstrated advanced development of chiral separation performances of CMSPs. Enantioselective monolithic columns for chiral separations by HPLC have many advantages over the conventional-packed and open-tubular capillary columns, such as higher separation, resolution, and the possibility of monoliths to avoid particle synthesis and the difficulty associated with discrete particles especially the need for end frits. The tendencies of organic-based polymer monolithic stationary phases to swell or shrink when exposed to various mobile phase solvents render the inorganic silica-based monolith a promising alternative. Micro- and nano-HPLC offer a rapid method for chiral separations and push the concept of green chemistry with reduced amounts of solvent, reduced sample sizes, and higher throughput for industry when compared with conventional HPLC. In this chapter, the phrases micro-HPLC and nano-HPLC were used interchangeably. The mechanism of immobilization of CSs in stationary phases varied in between particles as immobilization, functionalization, coating, and bonding to monolithic column or capillary. This is due to the lack of clear bold line for the manipulation and control in this field.

4.5   FUTURE PROSPECTIVE

By virtue of robust inclusion complexation and facile modification of compounds with selective functional groups, the development of novel CSs immobilized on monolithic matrices will continue to play more important roles in separation science and technology, with strong focuses on stable and achievable immobilization strategies, synthesis of novel CSs with multiple interaction sites, and combining CSs with each other. This should result in good stability and durability, and should provide cooperative and synergistic recognition sites for enhancement of recognition ability toward targeted analytes in both analytical and preparative scales. Ongoing research will focus on fast enantioseparation by the development of novel CS-based monolithic columns that will continue to dominate research interests in the coming future, which may be realized by immobilizing CSs onto the inner walls of microchannels. Immobilized CSs will likely play an important role for future pharmaceutical industry, drug discovery, food industry, development of agrochemicals, fragrances, and chiral pollutants.

From this chapter, we hope in the near future to have a well-defined and clear elucidation for different techniques used for adaptation of CS to the column. This is because many different phrases were used to describe the link between CS and column or capillary as functionalizing, coating, bonding, reacting, encapsulating, and immobilizing. Furthermore, it is very important to discriminate a clear and facile classification of monoliths according to type of CS, mechanism of binding between CS and monolith, or mechanism of separation of analytes.

REFERENCES

1. Aboul-Enein, H.Y and Wainer, I.W., *The Impact of Stereochemistry on Drug Development and Use*. 1997. New York, John Wiley & Sons.
2. Ali, I. and Aboul-Enein, H.Y., Introduction. *Chiral Pollutants: Distribution, Toxicity and Analysis by Chromatography and Capillary Electrophoresis*. 2004. New York, John Wiley & Sons, 1–35.
3. Younes, A.A., D. Mangelings, and Y. Vander Heyden, Chiral separations in reversed-phase liquid chromatography: Evaluation of several polysaccharide-based chiral stationary phases for a separation strategy update. *J Chromatogr A*, 2012. **1269**: 154–167.
4. Francotte, E.R., Enantioselective chromatography as a powerful alternative for the preparation of drug enantiomers. *J Chromatogr A*, 2001. **906**(1–2): 379–397.
5. Gil-Av, E. and B. Feibush, Resolution of enantiomers by gas liquid chromatography with optically active stationary phases. Separation on packed columns. *Tetrahedron Lett*, 1967. **8**(35): 3345–3347.
6. Han, S.M., Direct enantiomeric separations by high performance liquid chromatography using cyclodextrins. *Biomed Chromatogr*, 1997. **11**(5): 259–271.
7. Glajch, J.L., J.J. Kirkland, and J. Köhler, Effect of column degradation on the reversed-phase high-performance liquid chromatographic separation of peptides and proteins. *J Chromatogr A*, 1987. **384**: 81–90.
8. Wehrli, A. et al., Influence of organic bases on the stability and separation properties of reversed-phase chemically bonded silica gels. *J Chromatogr A*, 1978. **149**: 199–210.
9. McNeff, C. et al., Analytical advantages of highly stable stationary phases for reversed-phase LC. *LCGC North Am*, 2000. **18**(5): 514–529.
10. Cabrera, K., Applications of silica-based monolithic HPLC columns. *J Sep Sci*, 2004. **27**(10–11): 843–852.
11. Núñez, O., K. Nakanishi, and N. Tanaka, Preparation of monolithic silica columns for high-performance liquid chromatography. *J Chromatogr A*, 2008. **1191**(1–2): 231–252.
12. Kanatyeva, A.Y. et al., Monolithic stationary phases in liquid and gas chromatography. *Russ Chem Rev*, 2008. **77**(4): 393–400.
13. Wistuba, D., Chiral silica-based monoliths in chromatography and capillary electro-chromatography. *J Chromatogr A*, 2010. **1217**(7): 941–952.
14. Nakanishi, K. and N. Soga, Phase separation in silica sol-gel system containing poly-acrylic acid I. Gel formaation behavior and effect of solvent composition. *J Non-Cryst Solids*, 1992. **139**: 1–13.
15. Guiochon, G., Monolithic columns in high-performance liquid chromatography. *J Chromatogr A*, 2007. **1168**(1–2): 101–68; discussion 100.
16. Saunders, K.C. et al., Separation and sample pre-treatment in bioanalysis using mono-lithic phases: A review. *Anal Chim Acta*, 2009. **652**(1–2): 22–31.
17. Liao, J.-L., R. Zhang, and S. Hjertén, Continuous beds for standard and micro high-performance liquid chromatography. *J Chromatogr A*, 1991. **586**(1): 21–26.
18. Turson, M. et al., Monolithic poly(ethylhexyl methacrylate-*co*-ethylene dimethacry-late) column with restricted access layers prepared via reversible addition-fragmenta-tion chain transfer polymerization. *J Sep Sci*, 2011. **34**(2): 127–134.
19. Gibson, G.T.T., S.M. Mugo, and R.D. Oleschuk, Surface-mediated effects on porous polymer monolith formation within capillaries. *Polymer*, 2008. **49**(13–14): 3084–3090.
20. Ahmed, M. and A. Ghanem, Enantioselective nano liquid chromatographic separation of Racemic pharmaceuticals: A facile one-pot in situ preparation of lipase-based poly-mer monoliths in capillary format. *Chirality*, 2014. **26**(11): 754–763.
21. Peters, E.C., P. Metro, F. Svec, and J.M.J. Fréchet, Molded rigid polymer monoliths as separation media for capillary electrochromatography. *Anal Chem*, 1997. **69**(17): 3646–3649.

22. Svec, F., Porous polymer monoliths: Amazingly wide variety of techniques enabling their preparation. *J Chromatogr A*, 2010. **1217**(6): 902–924.
23. Nakanishi, K. et al., Structure design of double-pore silica and its application to HPLC. *J Sol-Gel Sci Technol*, 1998. **13**(1/3): 163–169.
24. Guiochon, G., Monolithic columns in high-performance liquid chromatography. *J Chromatogr A*, 2007. **1168**(1–2): 101–168.
25. Cabrera, K., G. Wieland, D. Lubda, K. Nakanishi, N. Soga, H. Minakuchi, K.K. Unger. SilicaROD—A new challenge in fast high-performance liquid chromatography separations. *TrAC Trends Anal Chem*, 1998. **17**(1): 50–53.
26. Puy, G. et al., Influence of the hydrothermal treatment on the chromatographic properties of monolithic silica capillaries for nano-liquid chromatography or capillary electrochromatography. *J Chromatogr A*, 2007. **1160**(1–2): 150–159.
27. Tanaka, N. et al., Monolithic silica columns for HPLC, micro-HPLC, and CEC. *J High Resolut Chromatogr*, 2000. **23**(1): 111–116.
28. Svec, F. and Y. Lv, Advances and recent trends in the field of monolithic columns for chromatography. *Anal Chem*, 2015. **87**(1): 250–273.
29. Colon, L.A. and L. Li, Organo-silica hybrid monolithic columns for liquid chromatography. *Adv Chromatogr*, 2008. **46**: 391–421.
30. Wu, M. et al., Preparation and application of organic-silica hybrid monolithic capillary columns. *Electrophoresis*, 2011. **32**(1): 105–115.
31. Stahlberg, J. et al., Structural basis for enantiomer binding and separation of a common beta-blocker: Crystal structure of cellobiohydrolase Cel7A with bound (S)-propranolol at 1.9 A resolution. *J Mol Biol*, 2001. **305**(1): 79–93.
32. Itoh, T. et al., Stereoselectivity and enantiomer-enantiomer interactions in the binding of ibuprofen to human serum albumin. *Chirality*, 1997. **9**(7): 643–649.
33. Bertucci, C. et al., Site I on human albumin: Differences in the binding of (R)- and (S)-warfarin. *Chirality*, 1999. **11**(9): 675–679.
34. Chuang, V.T.G. and M. Otagiri, Stereoselective binding of human serum albumin. *Chirality*, 2006. **18**(3): 159–166.
35. Hödl, H. et al., Chiral resolution of tryptophan derivatives by CE using canine serum albumin and bovine serum albumin as chiral selectors. *Electrophoresis*, 2006. **27**(23): 4755–4762.
36. Lämmerhofer, M., Chiral recognition by enantioselective liquid chromatography: Mechanisms and modern chiral stationary phases. *J Chromatogr A*, 2010. **1217**(6): 814–856.
37. Hong, T. et al., Preparation of graphene oxide-modified affinity capillary monoliths based on three types of amino donor for chiral separation and proteolysis. *J Chromatogr A*, 2016. **1456**: 249–256.
38. Fu, Y. et al., Bio-inspired enantioseparation for chiral compounds. *Chin J Chem Eng*, 2016. **24**(1): 31–38.
39. Jiang, T., R. Mallik, and D.S. Hage, Affinity monoliths for ultrafast immunoextraction. *Anal Chem*, 2005. **77**(8): 2362–2372.
40. Mallik, R. and D.S. Hage, Affinity monolith chromatography. *J Sep Sci*, 2006. **29**(12): 1686–1704.
41. Kalashnikova, I., N. Ivanova, and T. Tennikova, Macroporous monolithic layers as efficient 3-D microarrays for quantitative detection of virus-like particles. *Anal Chem*, 2007. **79**(14): 5173–5180.
42. Platonova, G.A. et al., Quantitative fast fractionation of a pool of polyclonal antibodies by immunoaffinity membrane chromatography. *J Chromatogr A*, 1999. **852**(1): 129–140.
43. Mallik, R., T. Jiang, and D.S. Hage, High-performance affinity monolith chromatography: Development and evaluation of human serum albumin columns. *Anal Chem*, 2004. **76**(23): 7013–7022.

44. Tetala, K.K.R. and T.A. van Beek, Bioaffinity chromatography on monolithic supports. *J Sep Sci*, 2010. **33**(3): 422–438.
45. Wang, P.G., Monolithic chromatography and its Modern Applications. 2010. St. Albans, UK, ILM publications.
46. Berruex, L.G., R. Freitag, and T.B. Tennikova, Comparison of antibody binding to immobilized group specific affinity ligands in high performance monolith affinity chromatography. *J Pharm Biomed Anal*, 2000. **24**(1): 95–104.
47. Gupalova, T.V. et al., Quantitative investigation of the affinity properties of different recombinant forms of protein G by means of high-performance monolithic chromatography. *J Chromatogr A*, 2002. **949**(1–2): 185–193.
48. Gustavsson, P.E. and P.O. Larsson, Continuous superporous agarose beds for chromatography and electrophoresis. *J Chromatogr A*, 1999. **832**(1–2): 29–39.
49. Calleri, E. et al., Development of a chromatographic bioreactor based on immobilized β-glucuronidase on monolithic support for the determination of dextromethorphan and dextrorphan in human urine. *J Pharm Biomed Anal*, 2004. **35**(5): 1179–1189.
50. Luo, Q. et al., High-performance affinity chromatography with immobilization of protein A and L-histidine on molded monolith. *Biotechnol Bioeng*, 2002. **80**(5): 481–489.
51. Pan, Z. et al., Protein A immobilized monolithic capillary column for affinity chromatography. *Analytica Chimica Acta*, 2002. **466**(1): 141–150.
52. Petro, M., F. Svec, and J.M.J. Fréchet, Immobilization of trypsin onto "molded" macroporous poly(glycidyl methacrylate-*co*-ethylene dimethacrylate) rods and use of the conjugates as bioreactors and for affinity chromatography. *Biotechnol Bioeng*, 2000. **49**(4): 355–363.
53. Xuan, H. and D.S. Hage, Immobilization of α1-acid glycoprotein for chromatographic studies of drug–protein binding. *Anal Biochem*, 2005. **346**(2): 300–310.
54. Martin del Valle, E., M.A. Galan Serrano, and R.L. Cerro, Use of ceramic monoliths as stationary phase in affinity chromatography. *Biotechnol Prog*, 2003. **19**(3): 921–927.
55. Yao, C. et al., High-performance affinity monolith chromatography for chiral separation and determination of enzyme kinetic constants. *Talanta*, 2010. **82**(4): 1332–1337.
56. Pfaunmiller, E.L. et al., Optimization of human serum albumin monoliths for chiral separations and high-performance affinity chromatography. *J Chromatogr A*, 2012. **1269**: 198–207.
57. Machtejevas, E. and A. Maruška, A new approach to human serum albumin chiral stationary phase synthesis and its use in capillary liquid chromatography and capillary electrochromatography. *J Sep Sci*, 2002. **25**(15–17): 1303–1309.
58. Mohammad, J., Y.M. Li, M. El-Ahmad, K. Nakazato, G. Pettersson, and S. Hjertéean, Chiral recognition chromatography of β-blockers on continuous polymer beds with immobilized cellulase as enantioselective protein. *Chirality*, 1993. **5**(6): 464–470.
59. Wistuba, D., Chiral silica-based monoliths in chromatography and capillary electrochromatography. *J Chromatogr A*, 2010. **1217**(7): 941–952.
60. Mallik, R. and D.S. Hage, Affinity monolith chromatography. *J Sep Sci*, 2006. **29**(12): 1686–704.
61. Mallik, R. and D.S. Hage, Development of an affinity silica monolith containing human serum albumin for chiral separations. *J Pharm Biomed Anal*, 2008. **46**(5): 820–830.
62. Mallik, R., C. Wa, and D.S. Hage, Development of sulfhydryl-reactive silica for protein immobilization in high-performance affinity chromatography. *Anal Chem*, 2007. **79**(4): 1411–1424.
63. Calleri, E. et al., Development of a bioreactor based on trypsin immobilized on monolithic support for the on-line digestion and identification of proteins. *J Chromatogr A*, 2004. **1045**(1–2): 99–109.

64. Xuan, H. and D.S. Hage, Immobilization of alpha(1)-acid glycoprotein for chromato-graphic studies of drug-protein binding. *Anal Biochem*, 2005. **346**(2): 300–10.
65. Schmid, M.G. et al., New particle-loaded monoliths for chiral capillary electrochro-matographic separation. *Electrophoresis*, 2004. **25**(18–19): 3195–203.
66. Mallik, R., H. Xuan, and D.S. Hage, Development of an affinity silica monolith con-taining alpha1-acid glycoprotein for chiral separations. *J Chromatogr A*, 2007. **1149**(2): 294–304.
67. Guillaume, Y.-C. and C. André, A novel chiral column for the HPLC separation of a series of dansyl amino and arylalkanoic acids. *Talanta*, 2008. **76**(5): 1261–1264.
68. Sancho, R. et al., Monolithic silica columns functionalized with substituted polypro-line-derived chiral selectors as chiral stationary phases for high-performance liquid chromatography. *J Sep Sci*, 2014. **37**(20): 2805–13.
69. Novell, A. and C. Minguillon, Monolithic silica columns with covalently attached octaproline chiral selector. Dependence of performance on derivatization degree and comparison with a bead-based analogue. *J Chromatogr A*, 2015. **1384**: 124–32.
70. Chen, Z., K. Uchiyama, and T. Hobo, Chemically modified chiral monolithic silica column prepared by a sol-gel process for enantiomeric separation by micro high-performance liquid chromatography. *J Chromatogr A*, 2002. **942**(1–2): 83–91.
71. Chen, Z. and T. Hobo, Chemically L-prolinamide-modified monolithic silica column for enantiomeric separation of dansyl amino acids and hydroxy acids by capillary elec-trochromatography and μ-high performance liquid chromatography. *Electrophoresis*, 2001. **22**(15): 3339–3346.
72. Schmid, M.G. et al., Fast chiral separation by ligand-exchange HPLC using a dynami-cally coated monolithic column. *J Sep Sci*, 2006. **29**(10): 1470–1475.
73. Schmid, M.G. et al., Enantioseparation by ligand-exchange using particle-loaded monoliths: Capillary-LC versus capillary electrochromatography. *J Biochem Biophys Methods*, 2007. **70**(1): 77–85.
74. Calleri, E. et al., Evaluation of a monolithic epoxy silica support for penicillin G acylase immobilization. *J Chromatogr A*, 2004. **1031**(1–2): 93–100.
75. Gotti, R. et al., Chiral capillary liquid chromatography based on penicillin G acyl-ase immobilized on monolithic epoxy silica column. *J Chromatogr A*, 2012. **1234**: 45–49.
76. Chankvetadze, B. et al., High-performance liquid chromatographic enantioseparations on monolithic silica columns containing a covalently attached 3,5-dimethylphenylcar-bamate derivative of cellulose. *J Chromatogr A*, 2004. **1042**(1–2): 55–60.
77. Chankvetadze, B., et al., High-performance liquid chromatographic enantioseparations on capillary columns containing crosslinked polysaccharide phenylcarbamate deriva-tives attached to monolithic silica. *J Sep Sci*, 2006. **29**(13): 1988–1995.
78. Ahmed, M., M. Gwairgi, and A. Ghanem, Conventional Chiralpak ID vs. capillary Chiralpak ID-3 amylose tris-(3-chlorophenylcarbamate)-based chiral stationary phase columns for the enantioselective HPLC separation of pharmaceutical racemates. *Chirality*, 2014. **26**(11): 677–682.
79. Fanali, C., S. Fanali, and B. Chankvetadze, HPLC separation of enantiomers of some flavanone derivatives using polysaccharide-based chiral selectors covalently immobi-lized on silica. *Chromatographia*, 2016. **79**(3–4): 119–124.
80. Ou, J. et al., Hybrid monolithic columns coated with cellulose tris(3,5-dimethylphe-nylcarbamate) for enantioseparations in capillary electrochromatography and capillary liquid chromatography. *J Chromatogr A*, 2012. **1269**: 372–378.
81. Rocco, A., A. Maruška, S. Fanali, Enantioseparation of drugs by means of continu-ous bed (monolithic) columns in nano-liquid chromatography. *Chemija*, 2012. **23**(4): 294–300.

82. Chankvetadze, B., Monolithic chiral stationary phases for liquid-phase enantiosepara-tion techniques. *J Sep Sci*, 2010. **33**(3): 305–314.
83. Guerrouache, M., M.C. Millot, and B. Carbonnier, Functionalization of macroporous organic polymer monolith based on succinimide ester reactivity for chiral capillary chromatography: A cyclodextrin click approach. *Macromol Rapid Commun*, 2009. **30**(2): 109–111.
84. Wang, H.-S., X.-Y. Feng, and J.-P. Wei, Biocompatible chiral monolithic stationary phase synthesized via atom transfer radical polymerization for high performance liquid chromatographic analysis. *J Chromatogr A*, 2015. **1409**: 132–137.
85. Guo, J. et al., Influence of the linking spacer length and type on the enantioseparation ability of β-cyclodextrin functionalized monoliths. *Talanta*, 2016. **152**: 259–268.
86. Lv, Y. et al., Preparation of novel β-cyclodextrin functionalized monolith and its appli-cation in chiral separation. *J Chromatogr B*, 2010. **878**(26): 2461–2464.
87. Ahmed, M. and A. Ghanem, Chiral beta-cyclodextrin functionalized polymer monolith for the direct enantioselective reversed phase nano liquid chromatographic separation of racemic pharmaceuticals. *J Chromatogr A*, 2014. **1345**: 115–27.
88. Ghanem, A. et al., Trimethyl-β-cyclodextrin-encapsulated monolithic capillary col-umns: Preparation, characterization and chiral nano-LC application. *Talanta*, 2017. **169**: 239–248.
89. Bayer, M., C. Hansel, and A. Mosandl, Enantiomer separation on monolithic silica HPLC columns using chemically bonded methylated and methylated/acetylated 6-*O*-*tert*-butyldimethylsilylated beta-cyclodextrin. *J Sep Sci*, 2006. **29**(11): 1561–1570.
90. Zhang, Z.B., M.H. Wu, R.A. Wu, J. Done, J.J. Ou, and H.F. Zou, Preparation of per-phenylcarbamoylated β-cyclodextrin-silica hybrid monolithic column with "one-pot" approach for enantioseparation by capillary liquid chromatography. *Anal Chem*, 2011. **83**: 3616.
91. Ghanem, A. et al., Immobilized beta-cyclodextrin-based silica vs polymer mono-liths for chiral nano liquid chromatographic separation of racemates. *Talanta*, 2015. **132**: 301–314.
92. Pittler, E. and M.G. Schmid, Enantioseparation of dansyl amino acids by HPLC on a monolithic column dynamically coated with a vancomycin derivative. *Biomed Chromatogr*, 2010. **24**(11): 1213–1219.
93. Guillaume, Y.C. and C. Andre, Fast enantioseparation by HPLC on a modified car-bon nanotube monolithic stationary phase with a pyrenyl aminoglycoside derivative. *Talanta*, 2013. **115**: 418–421.
94. Wu, H. et al., Enantioseparation of N-derivatized amino acids by micro-liquid chro-matography/laser induced fluorescence detection using quinidine-based monolithic columns. *J Pharm Biomed Anal*, 2016. **121**: 244–252.
95. Wang, Q. et al., Enantioseparation of N-derivatized amino acids by micro-liquid chromatography using carbamoylated quinidine functionalized monolithic stationary phase. *J Chromatogr A*, 2014. **1363**: 207–215.
96. Wang, Q. et al., Chiral separation of acidic compounds using an *O*-9-(*tert*-butylcarbamoyl)quinidine functionalized monolith in micro-liquid chromatography. *J Chromatogr A*, 2016. **1444**: 64–73.
97. Lubda, D. and W. Lindner, Monolithic silica columns with chemically bonded *tert*-butylcarbamoylquinine chiral anion-exchanger selector as a stationary phase for enantiomer separations. *J Chromatogr A*, 2004. **1036**(2): 135–143.
98. Zhang, C. et al., Macromolecular crowding-assisted fabrication of liquid-crystalline imprinted polymers. *Anal Bioanal Chem*, 2015. **407**(10): 2923–2931.
99. Chen, X. et al., In situ synthesis of monolithic molecularly imprinted stationary phases for liquid chromatographic enantioseparation of dibenzoyl tartaric acid enantiomers. *J Porous Mater*, 2011. **19**(5): 587–595.

100. Huang, Y.-P. et al., Preparation and characterization of a low-density imprinted monolithic column. *Chromatographia*, 2009. **70**(5–6): 691–698.
101. Bai, L.H. et al., Chiral separation of racemic mandelic acids by use of an ionic liquid-mediated imprinted monolith with a metal ion as self-assembly pivot. *Anal Bioanal Chem*, 2013. **405**(27): 8935–8943.
102. Ou, J. et al., Preparation and evaluation of a molecularly imprinted polymer derivatized silica monolithic column for capillary electrochromatography and capillary liquid chromatography. *Anal Chem*, 2007. **79**(2): 639–646.
103. Fairhurst, R.E. et al., A direct comparison of the performance of ground, beaded and silica-grafted MIPs in HPLC and turbulent flow chromatography applications. *Biosens Bioelectron*, 2004. **20**(6): 1098–1105.
104. Chiroli, V. et al., A chiral organocatalytic polymer-based monolithic reactor. *Green Chem*, 2014. **16**(5): 2798.
105. Preinerstorfer, B. et al., Enantioselective silica-based monoliths modified with a novel aminosulfonic acid-derived strong cation exchanger for electrically driven and pressure-driven capillary chromatography. *Electrophoresis*, 2008. **29**(8): 1626–1637.
106. Slater, M.D., J.M. Frechet, and F. Svec, In-column preparation of a brush-type chiral stationary phase using click chemistry and a silica monolith. *J Sep Sci*, 2009. **32**(1): 21–28.
107. Ghanem, A., T. Ikegami, and N. Tanaka, New silica monolith bonded chiral (R)-gamma butyrolactone for enantioselective micro high-performance liquid chromatography. *Chirality*, 2011. **23**(10): 887–890.

Index

Note: Page numbers followed by f and t refer to figures and tables, respectively.